Reading and Writing Strategies *for the* Secondary

SOCIAL STUDIES

Classroom *in a* PLC at Work®

Daniel M. Argentar
Katherine A. N. Gillies
Maureen M. Rubenstein
Brian R. Wise

EDITED BY
Mark Onuscheck
Jeanne Spiller

Solution Tree | Press

a division of
Solution Tree

555 North Morton Street
Bloomington, IN 47404
800.733.6786 (toll free) / 812.336.7700
FAX: 812.336.7790

email: info@SolutionTree.com
SolutionTree.com

Visit **go.SolutionTree.com/literacy** to download the free reproducibles in this book.

Printed in the United States of America

Library of Congress Cataloging-in-Publication Data

Names: Argentar, Daniel M., 1970- author. | Gillies, Katherine A. N.,
    author. | Rubenstein, Maureen M., author. | Wise, Brian R., author.
Title: Reading and writing strategies for the secondary social studies
    classroom in a PLC at work / Daniel M. Argentar, Katherine A.N. Gillies,
    Maureen M. Rubenstein, Brian R. Wise ; edited by Mark Onuscheck, Jeanne
    Spiller.
Description: Bloomington, IN : Solution Tree Press, 2021. | Includes
    bibliographical references and index.
Identifiers: LCCN 2020040029 (print) | LCCN 2020040030 (ebook) | ISBN
    9781949539035 (paperback) | ISBN 9781949539042 (ebook)
Subjects: LCSH: Social sciences--Study and teaching (Secondary) | Language
    arts--Correlation with content subjects. | Professional learning
    communities.
Classification: LCC H62 .A6629 2021 (print) | LCC H62 (ebook) | DDC
    300.71/2--dc23
LC record available at https://lccn.loc.gov/2020040029
LC ebook record available at https://lccn.loc.gov/2020040030

Solution Tree
Jeffrey C. Jones, CEO
Edmund M. Ackerman, President

Solution Tree Press
President and Publisher: Douglas M. Rife
Associate Publisher: Sarah Payne-Mills
Art Director: Rian Anderson
Managing Production Editor: Kendra Slayton
Senior Production Editor: Todd Brakke
Content Development Specialist: Amy Rubenstein
Copy Editor: Mark Hain
Proofreader: Jessi Finn
Cover and Text Designer: Abigail Bowen
Editorial Assistants: Sarah Ludwig and Elijah Oates

ACKNOWLEDGMENTS

Thank you to our significant others and families for supporting us through the process of writing this book. This book wouldn't be possible if we didn't have their continued support and encouragement. Thank you to Brandi, Ami, Nadav, and Alon Argentar; Nick, Cole, and Chase Gillies; Ryan, Camilla, and Kallie Rubenstein; and Erin, Colin, and Julianne Wise.

We are indebted to all of the mentors, teachers, and students who have shaped our thinking throughout our teaching careers. In particular, we are grateful for the partnership and collaborations we have enjoyed with the Adlai E. Stevenson High School communication arts department and D219 Niles Township High Schools. It is their commitment to literacy within our professional learning community that is the foundation for adapting and creating many of the strategies in this book.

Thanks go out to Adlai E. Stevenson High School social studies teachers Patrick Ambrose, Kristin Barrett, Nicholas Haan, Sarah Gutierrez, Kolleen Madeck, Paul Mazzuca, Naomi Pierce, Elizabeth Schienkopf, Kara Ward, Sandra Wright, and Marie Zlotnikov. Your passion, expertise, inspiration, and support have been invaluable.

Thank you to the Stevenson High School librarians, Toni Gorman and Jami Lopez.

Thank you to Niles North High School's many committed literacy teams and to the social studies teachers who have worked tirelessly to support their students through challenging text. It has been a pleasure working through the many challenges that are a part of this committed work.

Thank you to D219 Niles Township High Schools' literacy coaches and specialists Stephanie Iafrate and Mary Richards and to Scott Dahlberg, director and teacher of social studies. Your commitment to fostering a culture of literacy at D219 has tremendously impacted this important body of work.

We also thank the literacy specialists, coaches, and teachers from Downers Grove North High School and the Chicago Area Literacy Leaders (CALL) group for sharing their expertise and inspiring the important literacy work and collaborations we engage in daily.

Thank you to the administrative leaders at Adlai E. Stevenson High School and Niles North High School for their encouragement and support. By prioritizing literacy in these schools, they have allowed our work to grow and inspire our school communities.

Finally, a very special thanks to Mark Onuscheck, director of curriculum, instruction, and assessment at Adlai E. Stevenson High School, whose compassion, humor, inspiration, guidance, and friendship motivate and inspire us regularly. Without his steady leadership, there would be no book to write.

Solution Tree Press would like to thank the following reviewers:

Kristoffer Barikmo
Instructional Coach
Shawnee Mission East High School
Prairie Village, Kansas

Marcia Grimm
Spanish and Social Studies Teacher
Riceville Middle and High School
Riceville, Iowa

Nichelle Pinkney
Social Studies and World Languages
Program Coordinator
Klein Independent School District
Klein, Texas

Jared Walker
Social Studies Teacher
William Blount High School
Maryville, Tennessee

Visit **go.SolutionTree.com/literacy** to download the free reproducibles in this book.

TABLE OF CONTENTS

Reproducible pages are in italics.

ABOUT THE SERIES EDITORS

Mark Onuscheck is director of curriculum, instruction, and assessment at Adlai E. Stevenson High School in Lincolnshire, Illinois. He is a former English teacher and director of communication arts. As director of curriculum, instruction, and assessment, Mark works with academic divisions around professional learning, articulation, curricular and instructional revision, evaluation, assessment, social-emotional learning, technologies, and Common Core implementation. He is also an adjunct professor at DePaul University.

Mark was awarded the Quality Matters Star Rating for his work in online teaching. He helps to build curriculum and instructional practices for TimeLine Theatre's arts integration program for Chicago Public Schools. Additionally, he is a National Endowment for the Humanities' grant recipient and a member of the Association for Supervision and Curriculum Development, the National Council of Teachers of English, the International Literacy Association, and Learning Forward.

Mark earned a bachelor's degree in English and classical studies from Allegheny College and a master's degree in teaching English from the University of Pittsburgh.

Jeanne Spiller is assistant superintendent for teaching and learning for Kildeer Countryside Community Consolidated School District 96 in Buffalo Grove, Illinois. School District 96 is recognized on AllThingsPLC (www.AllThingsPLC.info) as one of only a small number of school districts where all schools in the district earn the distinction of model professional learning community (PLC). Jeanne's work focuses on standards-aligned instruction and assessment practices. She supports schools and districts across the United States to gain clarity about and implement the four critical questions of PLCs. She is passionate about collaborating with schools to develop systems for teaching and learning that keep the focus on student results and helping teachers determine how to approach instruction so that all students learn at high levels.

Jeanne received a 2014 Illinois Those Who Excel Award for significant contributions to the state's public and nonpublic elementary schools in administration. She is a graduate of the 2008 Learning Forward Academy, where she learned how to plan and implement professional learning that improves educator practice and increases student achievement. She has served as a classroom teacher, team leader, middle school administrator, and director of professional learning.

Jeanne earned a master's degree in educational teaching and leadership from Saint Xavier University, a master's degree in educational administration from Loyola University Chicago, and an educational administrative superintendent endorsement from Northern Illinois University.

To learn more about Jeanne's work, visit https://livingtheplclife.com, and follow @jeeneemarie on Twitter.

To book Mark Onuscheck or Jeanne Spiller for professional development, contact pd@SolutionTree.com.

ABOUT THE AUTHORS

Daniel M. Argentar is a literacy coach and communication arts teacher at Adlai E. Stevenson High School in Lincolnshire, Illinois. In his previous work as a sixth-grade teacher, he taught reading, language arts, social studies, and science. Since 2001, he has provided academic literacy support to struggling freshmen and sophomores, in addition to teaching college prep and accelerated English courses. In his coaching role, he partners with instructors from across all content areas to increase disciplinary literacy for students by running book studies, professional development sessions, and one-on-one coaching meetings.

Daniel received a bachelor's degree in speech communications from the University of Illinois at Urbana-Champaign, an English teaching degree and a master's degree in curriculum and instruction from Northeastern Illinois University, and a master's degree in reading from Concordia University Chicago.

To learn more about Daniel's work, follow @dargentar125 on Twitter.

Katherine A. N. Gillies works as a reading specialist and English teacher at Niles North High School in Skokie, Illinois, where she previously served as a literacy coach. Katherine serves as the lead architect of schoolwide literacy improvement work, including building a comprehensive system of intervention and support for struggling readers, as well as crafting research-based curricula to ensure the continued literacy growth for all students. Katherine leads several collaborative teams and cross-curricular initiatives aimed at using data to inform instruction, building capacity for disciplinary literacy, and employing responsible assessment practices in the secondary arena. She has presented on these topics at local and national confer-ences, including the National Council of Teachers of English.

Katherine earned a bachelor's degree in literature and secondary education from Saint Louis University, a master's degree in literacy, language, and culture with a reading specialist certification from the University of Illinois at Chicago, and a master's degree in educational leadership and administration from Concordia University Chicago. She is also a certified Project CRISS (Creating Independence through Student-owned Strategies) trainer.

To learn more about Katherine's work, follow @Literacyskills on Twitter.

Maureen M. Rubenstein is a literacy coach and special education instructor at Adlai E. Stevenson High School in Lincolnshire, Illinois. As a teacher, she works on individu-alized education plans with students who have diagnosed reading, writing, and emotional disabilities. In her coach-ing role, she partners with instructors from all content areas to work on disciplinary literacy. In addition to coaching individual teachers, she works with other literacy coaches to coordinate and implement book clubs, professional development sessions, and one-on-one coaching sessions.

Maureen received a bachelor's degree in special education from Illinois State University, a master's degree in language literacy and specialized instruction (read-ing specialist) degree from DePaul University, and a master's degree in educational leadership from Northern Illinois University. Maureen is also a certified Project CRISS (Creating Independence through Student-owned Strategies) instructor and certified to teach Wilson Reading.

To learn more about Maureen's work, follow @SHS_LiteracyMR on Twitter.

Brian R. Wise is a literacy coach and English department chair at Deerfield High School in Deerfield, Illinois. As a department chair, he leads the English department by facilitating professional development, supervision and evaluation, curriculum, and instruction. He has taught a wide array of English and literacy intervention courses throughout his teaching career. As a literacy coach for high school, he worked with faculty members from all content areas to build teachers' capacity for embedding literacy skills into classroom instruction and assessment.

Brian received his bachelor's degree in English education from Boston University, a master's degree in English from DePaul University, and master's degrees in reading and principal preparation from Concordia University Chicago.

To learn more about Brian's work, follow @Wise_Literacy on Twitter.

To book Daniel M. Argentar, Katherine A. N. Gillies, Maureen M. Rubenstein, or Brian R. Wise for professional development, contact pd@SolutionTree.com.

PREFACE

To begin this book, and to immediately demonstrate the value of professional learning communities (PLCs) in supporting positive, thoughtful collaboration, we want to share a real-life experience we had with a group of fellow teachers in our school in our role as literacy coaches. We believe this experience serves as an example of the familiar struggle occurring in many schools when teachers from various content areas strive to approach literacy instruction.

In the fall of 2015, we walked with confidence into our PLC collaborative experience with social studies teachers. Because social studies is about constructing the narrative of history, we felt confident that our expertise with English and literacy would make this work easy. We were wrong. We began our work by listening to our colleagues describe the various issues they encountered with their students. In truth, the list of issues was similar to discussions we had with other disciplinary teams, and we compiled the following list of reasons our social studies colleagues felt students struggled in their classes.

▶ Students might have weak historical knowledge, cultural knowledge, or both.

▶ Students might have poor knowledge of reading strategies, poor usage of reading strategies, or both.

▶ Students might have difficulty remembering and focusing on important details.

▶ Students might have difficulty making logical inferences.

▶ Students might function as pseudo-readers (Buehl, 2017) who fake their way through reading or avoid it altogether.

▶ Students might struggle with their core writing skills with a need to improve focus, increase use of evidence, and provide clearer justification.

In addition, we found that social studies presented two more challenging aspects for our work. First, when we examined the student makeup in most social studies classes, we noticed a uniquely wide variety of initial student literacy levels. Whereas many courses throughout the school were limited in age (one single grade level) and ability (regular or accelerated), social studies classes often included multiple ages and students of all learning styles and abilities. In fact, as we looked at some world history sections with that specific teacher team, each class seemed to have students with reading abilities ranging from sixth grade to post–high school. The team had their work cut out for them in supporting such a wide variety of students.

The other challenge we faced with our social studies team was the tension between content and instruction. Like other disciplinary teachers, there is real stress involved with covering so many units and topics during a semester or school year. The notion that teachers would have to explicitly include literacy skills in their instruction, potentially taking time away from content, met with understandable resistance. "Am I supposed to be a reading and writing teacher too?" we sometimes heard. Because we understood this concern, our challenge was to help our teachers recognize the long-term value of incorporating literacy-based strategies alongside disciplinary instruction to simultaneously expand students' content knowledge and skill development. We did not want our teachers thinking that teaching literacy skills would become a nuisance or that the effort wouldn't improve students' learning social studies content.

We recognize these concerns from the start because we want you to understand that literacy and social studies instruction merge directly with the three big ideas of a PLC: (1) we believe in integrating change that is *focused on every student's learning*, where teams systematically consider, implement, evaluate, and revise all changes; (2) we believe in supporting *collaboration* (a collective commitment) between experts seeking to solve an educational concern; and (3) we believe that we must focus on the *results* students produce and use that knowledge to adjust and improve instruction (DuFour, DuFour, Eaker, Many, & Mattos, 2016).

We dedicate this book to the literacy issues we think warrant teachers' attention when connecting literacy to social studies. We hope these ideas can help develop collaborative partnerships at your school, and we hope this book can serve as a strong resource for your teaching. Every teacher is a teacher of literacy.

Every Teacher Is a Literacy Teacher

This book is part of the *Every Teacher Is a Literacy Teacher* series, which provides guidance on literacy-focused instruction and classroom strategies for grades preK–12. The elementary segment of this series includes separate titles focused on instruction in grades preK–1, grades 2–3, and grades 4–5. While each of these books follows a similar approach and structure, the content and examples they include address the discrete demands of each grade band. The secondary-level books we've crafted for this series focus on how subject-area teachers in grades 6–12 need specific instructional strategies to approach literacy in varying and innovative ways. To address this need, we designed each secondary-level book to do the following.

▸ Recognize the role every teacher must play in supporting the literacy development of students in all subject areas throughout their schooling

▸ Provide commonly shared approaches to literacy that can help students develop stronger, more skillful habits of learning

▸ Demonstrate how teachers can and should adapt literacy skills to support specific subject areas

▸ Model how commitment to a PLC culture can support the innovative collaboration necessary to support the literacy growth and success of every student

▸ Focus on creating literacy-based strategies in ways that promote the development of students' critical-thinking skills in each academic area

You may immediately recognize how this approach differs from many traditional school practices and formats, where educators view literacy development as the job of English language arts (ELA) teachers, reading teachers, or teachers of English learners. It is often accepted practice that these teachers bear the responsibility of teaching skills such as vocabulary development, inferential skills, and writing. In many PLC cultures, we've seen collaboration generally begins by teaming teachers within like disciplines. Science teachers team with other science teachers, social studies teachers team with other social studies teachers, and so on. When teams form according to discipline, they tend to focus only on their content and discipline-based skills. Although these traditional approaches may be effective, we recognize the need for schools to support changes that make teaching literacy a responsibility for all teachers. We propose that schools use the power of their PLC to adopt the collective commitment that every teacher is a literacy teacher. Science teachers need to be literacy teachers. Mathematics teachers need to be literacy teachers. Social studies teachers need to be literacy teachers. World language and fine arts teachers need to be literacy teachers.

In this book, we build our case for the importance of literacy instruction in all content areas and emphasize how building collaboration among middle school and high school social studies teachers and literacy experts (specialists and coaches), in particular, is an impactful catalyst for supporting student growth. We also stress a strong commitment toward building instructional improvements that support the growth of every learner. When discipline-based teachers and literacy experts team up to connect literacy-based strategies with discipline-specific subject-area instruction, they build stronger approaches to instruction that improve learning (Stephens et al., 2011).

We recognize that many schools do not have dedicated literacy experts available to collaborate with social studies teachers around the challenges of building stronger readers and writers. To that end, we encourage you to use this book as a thought partner with your team or as your own personal literacy expert that can help you generate changes to support student learning. It will be a helpful companion as you deepen conversations and navigate choices that will positively affect student growth and development. It describes and gives examples of more than two dozen literacy-based strategies that you can integrate into the social studies classroom. You can use many of the strategies immediately; others require more preparation. In either case, we urge you to get started.

There are many reasons why social studies teachers in grades 6–12 need to be literacy teachers. Reading about social studies, writing about social studies, and thinking like a social studies expert require a mindset that focuses on elements of reading and writing that are fundamentally different from reading for pleasure.

Reading and writing about social studies require the following.

▸ A close attention to detail and cause-and-effect relationships

▸ An understanding of how details interconnect to build concepts significant to history and civics

▸ The ability to interpret data and understand patterns and trends

Literacy strategies create an infrastructure of supports that allow students to learn independently and confidently meet these criteria. Social studies teachers who focus on building stronger literacy strategies in their classrooms provide the necessary skills that support students' abilities to develop their own thoughts and opinions. Given that students have such a wide range of reading levels and a diverse range of background knowledge, it is all the more important to be successful in this endeavor. It requires your collaborative team to provide more literacy instruction along with the important social studies content that is part of your curriculum.

This introduction begins the journey by exploring the need for literacy instruction, establishing an understanding of disciplinary literacy, and detailing the scope of this book.

The Need for Literacy Instruction

Picture a student who is just beginning to read. What behaviors do you see as this student engages with text? What is he or she learning to do first? How does he or she grapple with the challenge of learning how to read? Chances are, you visualize this reader at the beginning stages, working to crack the alphabetic code—breaking apart and sounding out words, one syllable at a time, and likely dealing with simple language and colorful illustrations. The words the student is trying to read are already ones that he or she likely employs in conversation. This student is engaging in growing basic literacy skills: decoding, fluency, and automaticity. During this early phase of learning how to read, comprehension and meaning-making almost take a back seat to decoding. The reader is working on the mechanical process of learning to read.

As readers advance beyond the beginning stages of reading and grow in their abilities to read, they become more fully fluent and able to comprehend a text. They begin reading to learn rather than learning to read. At this point, the advanced reader possesses the ability to make meaning from what he or she reads—the process of reading is no longer dedicated to the mechanical process of encoding and decoding a text. Instead, it's dedicated to learning and thinking. More advanced readers can infer from and analyze what they read in a book, as well as what they encounter in the physical world, even when they have limited experience with a topic. Such readers possess the critical literacy skills they will need for college and success in the workplace. These critically literate students are ready to take on complex tasks and dive into disciplinary literacy tasks. Social studies students with advanced reading skills are ready to engage with historical, psychological, and sociological texts that deepen understanding. They can comprehend not only past events and gain insight into people and society but also the overarching story that binds all these together. This requires immense foundational background knowledge to allow for the kind of higher-level analysis that social studies disciplines require.

Now, what about the reader who is somewhere between these two phases—the reader who is neither a beginner nor advanced? What about the student who can encode and decode but struggles to apply this information to form new understandings? The reality that all teachers know and experience in their classrooms is that there are many students who fall into this place along the continuum, and there are many students who leave high school without the essential life skill of being critically literate. In fact, National Assessment of Educational Progress (NAEP) results detailed in *The Condition of Education 2018* report (McFarland et al., 2018) suggest that only 36 percent of eighth-grade students and 37 percent of twelfth-grade students in the United States possess literacy skills at or above the level of proficiency. Over 60 percent of these students have not met this readiness benchmark. This means that a majority of students are moving through middle school and high school without developing the literacy skills necessary to be successful in social studies classrooms and likely possess significant vocabulary and background knowledge deficits.

This is the group of students with which we are most concerned in this book. We know that this large group of students requires more attention and a higher concentration on skill development. Moreover, a specific portion of these students will continue to need support in even basic literacy skill development. It is this portion of our student population that seems to be the conundrum—often, we find these are the students who teacher teams struggle to support.

Unfortunately, the literacy struggles many students experience are not always transparent, even though they are all too familiar in classrooms. The graph in figure I.1 represents the increasing gap in literacy as students progress through their schooling, boldly demonstrating the challenges educators must work to solve.

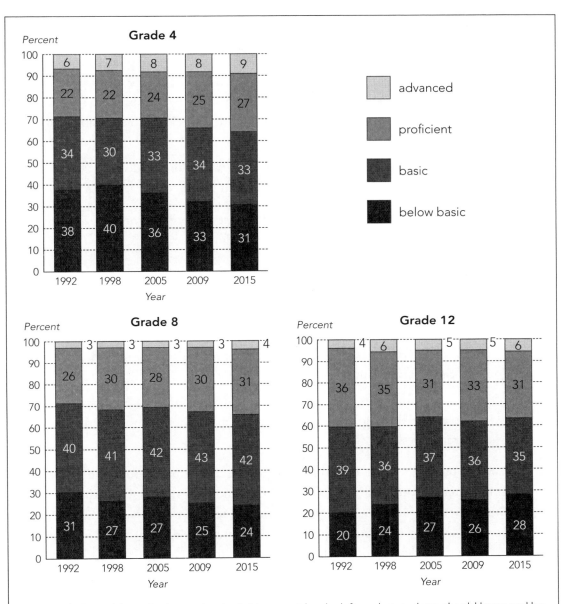

Note: Includes public and private schools. Achievement levels define what students should know and be able to do: Basic indicates partial mastery of fundamental skills, and Proficient indicates demonstrated competency over challenging subject matter. . . . Testing accommodations (such as extended time, small-group testing) for children with disabilities and English language learners were not permitted in 1992 and 1994. Although rounded numbers are displayed, the figures are based on unrounded estimates. Detail may not sum to totals because of rounding.

Sources: U.S. Department of Education, National Center for Education Statistics, & National Assessment of Educational Progress, n.d.

Figure I.1: Percentage distribution of fourth-, eighth-, and twelfth-grade students across NAEP reading achievement levels—1992–2015.

These students are in desperate need of instruction that cultivates the intermediate literacy skills that serve as a common foundation to disciplinary literacy. These skills include building academic vocabulary, self-monitoring comprehension, utilizing fix-it strategies to understand a text, and applying knowledge to a prompted task (Buehl, 2017). When all teachers acknowledge their responsibility to support literacy skills, collaborative teams can then shoulder the shared responsibility of student literacy and address these alarming statistics.

Research confirms a real need for disciplinary literacy instruction in the social studies classroom. Timothy and Cynthia Shanahan (2008) note the following.

- Adolescents in the first quarter of the 21st century read no better—and perhaps worse—than the generations before them.

- For many students, the rate of growth toward college readiness actually decreases as students move from eighth to twelfth grade.

- American fifteen-year-olds perform worse than their peers from fourteen other countries.

- Disciplinary literacy is an essential component of economic and social participation.

- Middle and high school students need ongoing literacy instruction because early childhood and elementary instruction do not correlate to later success.

Among the many concerns within collaborative discussions about teaching and learning, literacy continually ranks as one of the most worrisome. In many of our discussions with teachers across North America, teachers express three running concerns: (1) many students struggle with basic literacy skills, (2) many students read and write below grade level, and (3) many students do not know how to complete reading or writing assignments. We believe that teachers can utilize the power of their collaborative teams and the PLC process to bridge these gaps with a strong foundation of high-impact literacy instruction.

The gaps we see in student's literacy skills are staggering, and these gaps affect all areas of many students' education. As schools march students through their schooling, the statistics demonstrate that gaps in literacy increase throughout many students' elementary, middle, and high school years. Columbia University Teachers College (2005) reports many students find themselves reading three to six grade levels below their peers, many students struggle mightily to comprehend informational texts, and many students graduate from high school unprepared to enter a college-level experience. Columbia University Teachers College (2005) and Michael A. Rebell (2008) also present the following statistics, which present significant and long-standing concerns.

▸ By age three, children of professionals have vocabularies that are nearly 50 percent greater than those of working-class children, and twice as large as those of children whose families are on welfare.

▸ By the end of fourth grade, African American, Hispanic, and poor students of all races are two years behind their wealthier, predominantly white peers in reading and mathematics. By eighth grade, they have slipped three years behind, and by twelfth grade, four years behind.

▸ Only one in fifty Hispanic and African American seventeen-year-olds can read and gain information from a specialized text (such as the World section of the *New York Times*) compared to about one in twelve white students.

▸ By the end of high school, African American and Hispanic students' reading and mathematics skills are roughly the same as those of white students in the eighth grade.

▸ Among eighteen- to twenty-four-year-olds, about 90 percent of whites have either completed high school or earned a GED. Among African Americans, the rate is 81 percent; among Hispanics, 63 percent.

▸ African American students are only about 50 percent as likely, and Hispanics about 33 percent as likely, as white students to earn a bachelor's degree by age twenty-nine.

In its U.S. Adult Literacy Facts flier, ProLiteracy (n.d.) highlights the summation of this long-standing literacy crisis with the following facts detailing the reality of literacy in the United States and the catastrophic impact that illiteracy has on a multitude of social and economic factors. For example, more than thirty-six million U.S. adults cannot read, write, or do mathematics above a third-grade level. Forty-three percent of adults with low literacy levels live in poverty. When parents have low literacy levels, their children have a 72 percent chance of performing at the lowest reading level, receiving poor grades, developing behavior problems, having high school absentee problems, and dropping out. More than 1.2 million students drop out of high school each year (one out of every six). These jarring statistics undoubtedly reveal a systematic divide between those who are literate and those who are not, consequently deepening the inequities already present in our social structures. Visit https://bit.ly/3hhCdHG to read all of ProLiteracy's (n.d.) U.S. literacy facts, including statistics regarding English learners, unemployment, health literacy, and correctional facilities.

Statistical results like these are a stark reminder that teachers need to focus their attention on the literacy development of every student in their schools. For the grades 6–12 social studies teacher, the focus on developing students' abilities to comprehend informational texts should stand out as an important goal, as this goal is central to achieving social studies standards, developing skills, meeting expectations, and cultivating readers and writers who can approach a broad range of tasks independently and confidently.

As we note throughout this book, reading and writing strategies in a social studies classroom often mandate differing instructional approaches, requiring social studies teachers to collaborate in innovative ways so that they are meeting the needs of every student. Our literacy team's work finds a need to equip students with improved reading skills to address two groups of standards: (1) the College, Career, and Civic Life (C3) Framework for Social Studies (National Council for the Social Studies, 2017) and (2) the Common Core State Standards for English language arts (CCSS ELA). When social studies teams and literacy experts collaborate around these standards, they are sure to make a powerful impact on student learning.

Disciplinary Literacy

As your team gains confidence that students have a good grasp of basic foundational literacy skills, and as team members begin to see students develop more intermediate and advanced literacy skills, your team can move forward with tailoring literacy instruction with an eye toward disciplinary literacy. Even though students will need you to continue modeling the use of academic vocabulary and monitoring their comprehension, they will also be ready to attack complex texts— social studies texts that will challenge and engage students—with a disciplinary lens even as they practice building their skills.

For our purposes, a *discipline* is a unique expertise that schools often split into subject matter divisions such as mathematics, science, English, physical education, world languages, fine arts, and so on. Disciplinary literacy focuses on the literacy strategies tailored to a particular academic subject area. This book focuses on the expertise of social studies teachers who see the value of integrating specific literacy-building strategies into their classrooms due to the consistent employment of complex primary and secondary source texts. It's also worth pointing out that social studies is unique among academic disciplines in that this one content area includes multiple disciplines: history, economics, psychology, geography, civics, sociology, and so on. These each come with unique text conventions and writing expectations. Although most of the examples in this book emphasize history texts, we believe that social studies teachers focused on other disciplines can adapt these examples for their courses.

What would happen if we were to gather teachers from every discipline in a school and track the way they each address reading, writing, and speaking tasks? Predict how different content-area teachers would approach and work through literacy tasks. What similarities and differences would we observe among these varied disciplines?

thinking
BREAK

Thinking more broadly, because all teachers have unique expertise related to their individual academic fields, they often approach literacy-based tasks differently. Those differences stem from the diverse sets of expertise, interests, and background knowledge each professional brings to teaching and learning, and as a result, middle school or high school teachers often attend to literacy tasks differently based on that disciplinary expertise. After all, when a social studies teacher reads, writes, and speaks, he or she does so with certain goals and objectives in mind, such as determining the main idea, cause and effect, sequencing, and author's purpose, to name a few. While this may be a similar goal to other disciplines, what is specific to social studies is that the output of each of these skills represents the narrative nonfiction of the world and society. As such, from the perspective of citizenship, the stakes for comprehending social studies content are high.

There are also certain stylistic and conceptual norms professionals attend to in each discipline. A scientist, historian, businessperson, or any other professional addresses literacy tasks with norms and behaviors befitting his or her expertise and profession. That makes total sense; after all, each expert or professional has unique insider knowledge, with more background knowledge, subject-related vocabulary, and subject-related purpose than others without such dispositions. As a result, disciplinary outsiders often lack sufficient background knowledge and vocabulary to navigate a disciplinary text successfully. Literacy expert Doug Buehl (2017) suggests that the job of educators is to teach students how to think like they do—as disciplinary insiders. A social studies insider approaches reading tasks with specific goals and objectives that differ from insiders of other disciplines, with more attention to concerns such as comprehending text as a means to understand the past and to apply new understandings to the world around the reader.

Text comprehension in all disciplines generally follows a similar nine-step process, illustrated in figure I.2 (page 11). (See page 176 for a reproducible version of this process.) This figure and process derive from a guide Katherine Gillies (2019) developed when training peer tutors to help struggling readers navigate disciplinary texts; the figure provides a pathway for students to follow and teachers to model when comprehending a text. As we explore strategic comprehension steps and before, during, and after stages of reading in more detail throughout this book, we will demonstrate how application, connection, and extension of literacy skills unfold under the specific disciplinary lens of social studies.

Given the difference between disciplinary insiders and outsiders, it makes little sense for content-area teachers to instruct students to read and write with the same general strategies and moves in every content area. After all, if we know that each content area has its own thinking style, it makes sense that teachers support students in consuming and producing texts with the same unique thinking style required of each discipline. Even students who have a solid foundation of general strategies may struggle with the specific demands of disciplinary texts. Instead of using generic strategies in every class and across the school, providing students with a varied strategy toolbox to meet disciplinary demands better equips them as disciplinary insiders to read like scientists, mathematicians, historians, and so on (Gabriel & Wenz, 2017).

Over time, our literacy team has made positive strides toward building disciplinary literacy strategies that support learning in more directed, focused, and attentive ways. We've learned that we should apply more specific strategies to different disciplines in ways that help support learning. When we speak of this shift to disciplinary literacy and training students to be insiders, we intend to teach students to think differently in each classroom they encounter during their day. Note that in Collaboration Around Consistent Language (page 134), we encourage schools to establish a common literacy vocabulary for teachers to use across the school. We believe this helps students to build a general understanding of common terms such as *claim*, *evidence*, and *reasoning*. Through this common use of language, students can build on the foundational knowledge of these literacy concepts in discipline-specific ways as teachers become more comfortable teaching students how to read, write, and think like experts in their classroom. This is the goal of disciplinary literacy and why we often ask teachers who wonder how to teach a text, "How would you, as an expert, address the task?" As they think through their

Did I . . . ?	Strategic Comprehension Step	Before, During, or After Reading
☐	**Preview text**, ask questions, and make predictions.	**Before:** Focus and get ready to read.
☐	**Recall** what you already know about the topic.	
☐	Set a **purpose** for reading.	
☐	Make a **notetaking plan** for remembering what's important.	
☐	Define **key concepts** and important vocabulary whenever possible.	
☐	Keep your **purpose** for reading in mind.	**During:** Stay mentally active.
☐	**Make meaning** by: • Asking questions • Putting the main ideas into your own words • Visualizing what you read • Making notes to remember what's important • Making connections between the text and people, places, things, or ideas	
☐	**Be aware** of what's happening in your mind as you read. Consider: • Am I focused or distracted? • Do I need to go back to a part I didn't get and reread it? • What are my reactions to what I am reading?	
☐	**Reflect on** what you've read. Consider: • Did I find out what I needed or wanted to know? • Can I summarize the main ideas and important details in my own words? • Can I apply what I have learned? • Can I talk about or write about what I have learned?	**After:** Check for understanding.

Source: © 2019 Katherine Gillies. Adapted with permission.

Figure I.2: Reading-comprehension process poster.

own processes, often a strategy or a focus emerges that is unique to their discipline. This allows us to help teachers recognize the value of thinking about their discipline in relation to literacy, which is particularly important in the teaching of social studies, as students work to gain insights into the texts they read, weigh and value a text's details, and develop their capacity to select evidence, compare and contrast, synthesize, and critically analyze information.

About This Book

Our goal for this book is to support collaborative partnerships in schools to address social studies teachers' literacy concerns and better equip them for their work in social studies classrooms. Partnerships may occur within social studies teacher teams, between teams and literacy experts, or between different teacher teams throughout the school. We aim to connect that work with literacy strategies to develop students' understanding and skills as they read and write about social studies and learn to think like historians, sociologists, economists, psychologists, anthropologists, or other social studies insiders. Collaboration on these strategies will create new ways to heighten students' abilities to approach more complex texts with confidence and advance their abilities to think critically. The following sections explore the scope of this book, the common language we use throughout, and the individual chapter contents.

Scope

We designed this book to help literacy leaders collaborate and build literacy capacities in middle school and high school environments. We talk more specifically about what makes a literacy leader in the next section, but what's important to know here is that anyone in your building who takes on the mantle for driving literacy advancement in the classroom is a literacy leader. In elementary school, teachers work hard to teach students to *learn to read*. In middle school and high school, the goal is to teach students to *read to learn*. There is a big difference between the two approaches, as we noted earlier (see The Need for Literacy Instruction, page 3). Moreover, social studies teachers want students to learn from the reading and writing tasks the ability to think critically about the reading and writing they do.

As your team works to approach these challenges, team members need to recognize that each school is unique and each student is unique: there is no one-size-fits-all pathway to literacy development. Within this book, there is a continuum of supports related to the varying needs of each school and each classroom. Sometimes teachers might require short-term, immediate literacy triage; sometimes long-term, sustained collaborative development between team members is necessary; or sometimes there is a need for *both* triage and sustained literacy-based professional development. We recognize that strong, consistently applied literacy strategies can and will help all readers develop their potential, so we invite you to adapt the strategies we offer in this book to your unique needs. Many of the same literacy strategies that work for less complex literacy tasks still apply to more complex tasks—the only difference is the difficulty level. The skills that students need to apply remain the same and, with consistent application, become ingrained habits of the mind. As social studies teams collaborate on their work, staying committed to literacy-based strategies will help all students advance.

Common Language

For the purposes of this book, and to avoid getting confused by education jargon, we recognize that we need to have a common understanding of literacy and a common language around literacy development. For instance, we use the word *text* to mean a reading, an article, a chart, a diagram, a cartoon, a media artifact, and so on. There are many texts teachers ask students to read, and they can be in many formats. In addition, the term *literacy leader* can apply to a variety of educational roles. Throughout the book, a literacy leader can be anyone in your building, such as an administrator, teacher leader, reading specialist, or literacy coach. A literacy leader is someone who has a knowledge base in literacy and wants to improve the overall literacy skills of a school environment or institution. If you don't have a literacy leader at your school, don't let that stop you. Remember, you can use this book as a thought partner and become your school's literacy leader. The goal is to get started with the demanding challenges of literacy that need to be tackled now, with or without a literacy coach or a preexisting school literacy leader championing the work. Any teacher and team of teachers can initiate the changes that are necessary to support student learning; we mean for this book to help guide your understanding of how to approach these changes in teaching practices.

In this book, you will frequently see the abbreviation *PLC*, which is "an ongoing process in which educators work collaboratively in recurring cycles of collective inquiry and action research to achieve better results for the students they serve" (DuFour et al., 2016, p. 10). As we established in the preface to this book (page xv), a PLC consists of a schoolwide or districtwide culture of *collaboration* that uses a *results orientation* to achieve *learning for all*. We believe that a commitment to these three big ideas can help to support and innovate literacy in every classroom, and we believe that PLC culture promotes changes that will effectively support all students.

Within a PLC culture, collaborative teams meet on a consistent basis to build innovative practices concentrated on student growth and learning. We use the term *team* throughout the book with the understanding that all teams are interdependent and are professionally committed to continuous improvement. We know that teams may look different from building to building, and we know that schools need to configure teams differently based on building resources. In this book, we use *teams* generically to refer to social studies teams who are collaborating in focused ways to address literacy concerns for student learning in their classrooms. We also recognize that you might be a team of one—a singleton instructor who teaches an elective or is the only person teaching a grade-level course. For those singletons, we encourage you to be creative in finding ways to collaborate and discuss how to make use of literacy strategies more effectively. Reach out to teachers of different grade levels or in other area schools. Or, use the community forum at the AllThingsPLC website (www.allthingsplc.com) to locate a content-area teacher to collaborate with. There is great value in collaborating around how to use a strategy or make it more effective for your specific students, and in a PLC, every educator must be part of a collaborative team.

Chapter Contents

In chapter 1, we lay out fundamental aspects of the collaborative work teams engage in to address teaching literacy in the social studies classroom. In chapter 2, we begin with more in-depth discussions about foundational literacy and many immediate interventions for literacy difficulties that require a fast solution. We call this *literacy triage*. From there, we focus on disciplinary literacy collaboration for prereading, during reading, and postreading in chapters 3 through 5, respectively. Chapter 6 then offers guidance for teaching writing in the social studies classroom. Within these chapters, we intentionally focus on supporting a deeper, focused approach to literacy instruction within social studies. We offer classroom strategies that are the result of collaborative explorations by literacy leaders and content-area teachers, providing clarity on how varying perspectives inform instruction. For each example, we discuss that particular strategy's purpose, application, and literacy focus. We also indicate how teams can modify each strategy to support students who are struggling (including students who qualify for special education or who are learning English) or to extend learning for students showing proficiency. Finally, chapter 7 covers ideas for formative and summative assessment and feedback.

Throughout this text, there are opportunities called Thinking Breaks. We intend for these to help you reflect on current practices, challenges, and opportunities for growth in working with literacy in the social studies classroom. We know that you might do this naturally, but these are the points where we think it is important to slow down and consider ways to apply the strategies we suggest to your own students. In addition, each chapter ends with a series of questions to encourage thinking about your team's collaborative considerations. These questions highlight ways for teams to discuss, collaborate on, or implement disciplinary literacy ideas. Your team can use these discussions to build more directed literacy practices as you target your specific grade-level curriculum.

Ultimately, we hope this book and series are not only resources for ideas you can implement immediately in the classroom but also sources of inspiration for collaborative opportunities between literacy experts, leaders, and content-area instructors to increase literacy capacity in your building or buildings.

As I am reading and using this book as a resource to support my teaching, what do I want to get out of the content?

Note these three considerations for your team: (1) use this book as a book study, (2) break the book down chapter by chapter and focus on specific changes, and (3) prioritize your concerns for student learning and how to best support the literacy development of your social studies students.

thinking
BREAK

Wrapping Up

Building collaborative teams focused on literacy development can be challenging. We know you are extremely busy and have enormous amounts of content to cover, so you may be reluctant to add another layer to your already demanding workload. However, given the NAEP data that show more than half of U.S. twelfth graders graduate high school without preparation for advanced critical thinking (McFarland et al., 2018), we must pause and consider what we are all doing as educators to better prepare students for the future. Providing students with intermediate literacy and disciplinary literacy skills is an important step toward building the advanced literacy proficiency they will need throughout their educational careers and beyond.

Collaborative Considerations *for* Teams

- What are some of the unique features of texts used in social studies classrooms, including history, economics, psychology, and so on?

- What kind of criteria do experts in your discipline look for when they read?

- What deficiencies do you notice in your students that might obstruct their understanding of your content?

- How might your team provide experiences and vocabulary to help students feel more confident in reading texts used in a social studies classroom?

CHAPTER 1

Collaboration, Learning, and Results

Whose job is it to teach literacy skills to students, both struggling and striving? Social studies is a text-heavy content area, but most social studies teachers are disciplinary experts with degrees in history or other social studies fields, not reading endorsements or specialist literacy credentials. Most likely, even their teacher-preparation programs didn't provide extensive coverage of literacy instruction. As a result, some team members may assume this responsibility lies with ELA teachers or, in the case of some schools, a specific reading department. While it is true that these teachers must take responsibility for explicitly teaching students how to navigate text and providing scaffolds and supports so that students' literacy abilities will grow, the same is also true for teachers of all content areas. Social studies teachers cannot assume that the English department is solely responsible for the cultivation of students' literacy. The reality is that, while reading teachers are skilled in methods of research-based instructional intervention and support, one class period a day is not enough to stretch and grow students' skills to where they need to be for post–high school life.

Just like any other disciplinary team, a social studies team dedicated to improving disciplinary literacy needs to ask difficult questions about curriculum and teaching practices. Social studies teachers are a busy bunch; the vast range of topics explored within a single course (to say nothing of eras in a world history class) can be overwhelming, making it easy for social studies teachers to fall into the role of daily storyteller and text summarizer for students. However, identifying common concerns as a team almost always leads to an opportunity to identify a solution-focused literacy need. This is your team's opportunity to include literacy in the collaboration process around instruction.

Across North America, PLC schools are making a collective commitment to the power of collaboration. They are tackling long-standing concerns in education by bringing together leadership and teacher teams to make stronger curricular and instructional choices, and they are getting better and better at making use of assessment practices that support the formative development of all students (DuFour et al., 2016). Indeed, collaboration plays a crucial role in the success of any school dedicated to building a PLC culture. When teams use collaborative time wisely—when they adopt action steps that are clearly designed, intentional, and focused—new and innovative ideas emerge.

This chapter helps you identify how to initiate collaboration around literacy by applying PLC fundamentals and build teacher teams within your school to support meaningful teamwork that leads to student growth and reflective teaching practices. We offer guidance for teams and leaders and examine how to approach meeting logistics. We then examine in detail the work teams carry out in collaborative meetings, including analyzing standards, setting goals, identifying students' existing literacy skills and needs, and finding connections between the content area and literacy skills.

Collaboration Within a PLC

PLCs are a pivotal force for progress in schools, as they are all committed to three big ideas: (1) a focus on learning, (2) a collaborative culture, and (3) a results orientation (DuFour et al., 2016). Within our literacy work with social studies teachers, we keep these three big ideas at the core of our efforts, and we direct our commitment to literacy in every content area by continuously addressing the four critical questions of a PLC.

1. What knowledge, skills, and dispositions should every student acquire as a result of this unit, this course, or this grade level?

2. How will we know when each student has acquired the essential knowledge and skills?

3. How will we respond when some students do not learn?

4. How will we extend learning for students who are already proficient? (DuFour et al., 2016, p. 36)

We recognize that teams are configured in varying ways depending on the school. For instance, you might have curriculum teams, grade-level teams, content-focused teams, or teams comprised of singletons (the lone teacher for a grade level or content area in a school). No matter how your teams are structured, when working toward integrating literacy-based strategies, we hope they have the means to collaborate with a literacy expert in your building, such as a reading expert, specialist, or coach, who can provide insight into varied and supportive literacy strategies that your team must integrate into its instructional practices. As we wrote in the introduction, though, if your building does not have a literacy expert, we designed this book also to function as a thought-provoking substitute that will guide you in generating ideas and making actionable plans.

While the size and scope of work in a team can differ greatly from one context to another, many teams' initial focus is often a specific, discrete task that later evolves into more layered tasks and discussion topics. Looking at teams as groups of professionals who come together to collaborate, PLC architect Richard DuFour (2004) notes successful teams:

▸ Have common time for collaboration on a regular basis

▸ Build buy-in toward a discrete and overarching common goal

▸ Build a sense of community

▸ Engage in long-term work that continues from year to year

▸ Evolve—but do not completely change—membership each year

In addition to these characteristics, we find it helpful to have a team facilitator or point person who creates agendas and monitors discussion. It is also important to encourage open, honest, discussion-based dialogue that respects and includes all ideas concerning student learning. We recommend teams find opportunities to reach out to and include any colleagues whose work might overlap with that of the team's. When focusing on literacy, this might include an ELA teacher, reading specialist, or literacy coach; however, it might also include reaching out to teachers who specifically support a student in a team member's classroom, such as an intervention specialist, an arts teacher, or an English language teacher. Remember, collaborative team meetings work better when they are focused on actionable items that will serve to extend the professional learning of the teachers. The following sections go into more detail on the configuration of teams, the role of leaders within a team, and the logistics for team meetings.

Team Configuration

We want to make a few explicit recommendations for configuring teams effectively when working on literacy-based strategies. There are a number of different approaches to establishing a team, and resources vary in many schools. Sometimes teachers need to conceive of teams differently when thinking about how to make changes that will add more literacy expertise to the team dynamic. Here are a few considerations when building teams focused on literacy.

1. Ideally, incorporate a literacy expert on a curriculum- or content-focused team to serve as an informed thought partner. An individual with literacy expertise can help support instructional changes with a greater variety of approaches and can also help select strategies that are better aligned to course outcomes. For purposes of the social studies classroom, the literacy expert can help adapt strategies focused on critical-thinking areas, such as building background knowledge through contextualized vocabulary development, understanding a text's main idea, and providing strategies to explore a text's details.

2. Some teams work *horizontally*, meaning they collaborate within their specific grade level or within a particular content-based course with many sections and many teachers. Some teams work *vertically*, where they meet with teachers from different grade levels to ensure curricula are interconnected and function to build learning year to year. When considering literacy-based strategies, consider how your team is constructed and your goals for integrating literacy-based strategies. Why and how should your team apply strategies to support students throughout a particular grade level or within a social studies course that many teachers instruct? When and why should a team use a strategy? If working vertically, how and why might a team of instructors teach and reteach strategies year after year while ensuring students reach the next course or grade level with the essential knowledge they will need? How do these strategies become habits over the sequence of courses and throughout grade levels?

3. Some teams will not include or have access to a literacy expert. In these cases, consider what other resources can serve as a strong, literacy-informed, and reflective collaborative partner. Consider the expertise that might come from the ELA department. Is there a reading or writing teacher available to come to your team meetings? If not in your building, is there a literacy expert within your district—an expert in an elementary,

middle, or high school building—who might be able to help? Is there expertise in the special education department or an intervention specialist with a background in the area of literacy? Seek to use resources available to you.

4. Consider how your team might make collaborative use of this book in more directed ways. For instance, is there a specific literacy skill where students are struggling? Consider focusing on one aspect of social studies reading, such as prereading, that the team wants to improve. Or, if you are a singleton teacher, how might this book be a resource to help you reflect on your current teaching practices and grow your learning? To improve, teams often need to seek outside resources that do not currently exist within the schools they are working in. In these cases, widen your definition of a team and consider the larger community of educators that is willing and ready to help other teachers to learn. Reach out to professional organizations, attend conferences, or utilize the tools at AllThingsPLC (www.allthingsplc.com) that can help connect your team to discussions that will influence positive changes.

Leaders' Roles

Collaboration is at the crux of literacy work and is truly an essential component of professional growth. While this may seem obvious, there are several reasons achieving authentic collaboration can be a challenge. As discussed previously, we commonly hear "There isn't time" as a primary roadblock. Consequently, the first step in the collaboration process takes place between literacy leaders and administrators who need to work together to find time to build effective, supportive changes for how teams approach literacy in the social studies classroom.

Your past experiences with using immediate strategies to address students' most critical literacy needs provide an opportunity to approach your administrators and create a long-term plan for schoolwide literacy. The following chapters provide information and examples of how to begin this process, but here are some starting points to consider when communicating with administration and when engaging with the upcoming chapters in this book.

▸ Identify the immediate problem and show the evidence of the problem with data.

▸ Identify your ultimate goal for the students in your social studies classroom. Know what you want to change.

▶ Provide a list of strategies you have tried in your classroom. Identify the ways you and your team have worked to build change.

▶ Provide suggestions for resources or ideas that can help you and your team accomplish your goals and get your students to a more successful place. Collaborate with your administrators over how your team can accomplish these goals, with the needed action steps and the right supports in place.

Responding to your school's literacy needs cannot occur through an occasional meeting or a purchased program. Everyone needs to be on board with the challenging work of moving students' literacy competencies in the right direction, and leadership must support and respect this collaboration to ensure each teacher is open to his or her own education and professional growth. Having said that, we want to offer a few tips for literacy leaders engaged in this work.

If you are the literacy leader in your school, consider yourself a host to other teaching teams. You want the collaboration between you and the teams you work with to be as positive as possible so everyone can be an effective participant. Often as a leader in literacy, you will need to serve as the glue that holds the pieces together. You will need to develop the agendas and send out the invites and reminders. Your personal goal is to keep this team moving forward and committed to literacy-based strategies. Be authentic: you are not the person with all the answers, and the tone you adopt must be collaborative rather than hierarchical. The credit for the good work that comes from the focused collaborative team goes to the committee as a whole and not any one individual or leader.

thinking
BREAK

What additional questions or thoughts could you consider when approaching administrators about creating a schoolwide literacy program that can specifically assist in the social studies classroom?

Collaborative Meeting Logistics

Based on our experiences, there are four simple logistical action steps you can take that will help ensure fluid and timely collaboration.

1. **Work creatively and collaboratively to carve out a common regular time and space for your team to meet:** From our experiences, carving

out common time is the first step to developing a high-functioning team in a PLC. While a progressive education model allows for consistent and regular professional collaboration during the school day (due to the complex nature of the secondary school day), many schools do not have a schedule that reflects this. While a regular weekly team block where students start school late or leave school early is often the ideal way to ensure teacher teams can meet in focused ways, we also recognize that many school districts are still working toward supporting structures that empower PLC cultures. If your school has not yet set up a structured time for teacher teams to meet, we suggest planning ahead to ensure team meetings allow for collaboration about literacy. You might need to work with an administrator who will help support the collaborative time necessary to innovate positive changes. For example, you might ask for release periods or release days throughout the year so you can accomplish your team's goals.

2. **Create a regular meeting schedule, and make sure everyone knows the plan:** Collaborative teams must ensure they dedicate their meeting time to fostering the commitments of a continuously developing PLC. Early on, establishing norms that set focused, actionable goals helps teams achieve their purpose and helps to establish commitments. We suggest using SMART (specific and strategic, measurable, attainable, results oriented, and time bound) goals as a guiding tool (Conzemius & O'Neill, 2014). By setting up SMART goals, your team is more likely to stay focused, learn, and be driven to succeed. As literacy coaches, we know that time is precious, and when our social studies teachers meet with us, we know that our collaboration needs to have purposeful, specific outcomes. In building SMART goals toward literacy, we encourage taking the time to create an action-driven schedule that is paced, practical, and respectful of everyone's many different commitments. What is realistic depends on your structure and your team's purpose. From our personal experiences, we think that meeting less than every other week often means that team members are unable to prioritize making changes in literacy. Team meetings that are too infrequent mainly consist of recapping what happened at the last meeting and miss the mark on cultivating productive collaboration that leads to changes to literacy practices in the classroom. To ensure all team members are aware of upcoming meetings, create electronic calendar invites, send out a paper copy of dates and plans that

members can post at their desks, and send reminders. Always attach your agenda to encourage thoughtful preparation prior to the meeting.

3. **Identify and use a consistent meeting location:** We encourage finding a consistent space for your team to meet. It is counterproductive when people are always searching for a changing location. Ideally, this space will be free of other distractions, comfortable, and well equipped for the work you will be doing; for example, have access to a whiteboard, projector, laptop, or other devices.

4. **Create digital files that capture agendas and notes:** From the beginning, create an ongoing digital hub for agendas and notes—use tools like Slack, Google Docs with Google Drive, or Microsoft OneNote with Microsoft OneDrive. Ideally, you want something that allows all members to contribute independently and ensure full, online access to your materials outside of your collaboration time.

thinking
BREAK

○ Do you see any natural opportunities within your daily schedule for team time?

○ Are there any predictable patterns you notice in your department, school calendar, or schedule that would allow for meeting time?

Standards Analysis and Goal Setting

As we noted before, strong teams agree on what they want all students to learn, thus addressing the first critical question of a PLC. While your team may have a general overarching literacy goal right out of the gate, it must still dedicate time to the specific development of discrete goals that identify what it wants students to learn. Often, it is necessary to do some research to establish an effective goal. Before defining any sort of common goal for student learning, spend time examining your school's social studies curriculum standards and assess the current implementation of these standards with a productive and critical eye. Your team must unpack the social studies content standards and identify the literacy skills they require for mastery.

Another way to think of *content standards* is to frame them as desired student outcomes—the concepts that you want your students to have mastered at various checkpoints throughout your social studies curriculum. Your team will *unpack* (or *unwrap*) these standards to determine the skills necessary to achieve them. These building blocks represent the *process standards* (or learning targets) that will help your students reach the content standards; they are essentially the vehicle to transport students from *novice* to *skilled* on the mastery continuum. Process standards are comprised of a variety of focused tasks that teachers support via carefully sequenced literacy strategies used during instruction. For example, teaching students to read for details that help them fully understand the circumstances that led to a historical event is essential for their clear understanding.

Investigating current practices and detailing desired outcomes require your team to have thoughtful conversations about the curriculum standards in your social studies department. This work includes answering the following questions.

▸ What standards will we prioritize?

▸ What skills will we focus on that support learning these standards?

▸ What are our expectations for student performance?

▸ What are the criteria we will use to evaluate and support students' learning?

Answering these questions means your team must unpack the wordy and lengthy verbiage of your social studies content standards and identify what matters most to your specific course or context. In other words, your team needs to determine which standards within the curriculum it will prioritize for instruction. The standards that are most vital for students to learn become your *power standards* (sometimes called *priority standards* or *essential standards*; Crawford, 2011; DuFour et al., 2016). As part of this process, it is helpful to detail your team's understanding and the department's approach to teaching the standard as well as the evidence of student learning your team will collect to demonstrate that students have achieved these skills. Use figure 1.1 (page 26) to discuss with your team the content standards you believe are power standards and the evidence of learning that you expect students to demonstrate. This process ensures your team addresses the first two critical questions of a PLC.

We urge you not to overlook the steps in the reproducible "Content-Standard Analysis Tool" (page 117). While it is true that content standards may shift over time, it is also critical that teams continue to evolve curriculum alongside changing standards. While it is tempting to jump to identifying power standards, make sure that you don't skip the unpacking of *all* standards first. Curriculum teams that overlook non-power standards often fall into the trap of simply overstating core

| Content Standard | Artifacts | | Power Standard ☐ |
	Lesson Plans and Course Materials	Student-Generated Evidence of Standard	
Unpacked (process standards):			Opportunities and Next Steps
	Strengths and Needs	Strengths and Needs	

Content Standard-Analysis Steps (Use this with your team to guide collaborative discussions.)

1. **Unpack:** What is the standard really saying? Put the standard in your own words and determine the individual process standards inherent within the content standard.

2. **Identify artifacts:**
   - Lesson plans and course materials—
     – How is the standard being taught?
   - Student-generated evidence—
     – How are students demonstrating mastery?

3. **Assess the strengths and needs of these artifacts:** Where is there room for improvement?

4. **Identify power standards:** Which content outcomes are the most essential to your content area?

5. **Identify next steps:** As your team's work continues, identify potential teaching and learning opportunities.

Figure 1.1: Content-standard analysis tool.

*Visit **go.SolutionTree.com/literacy** for a free reproducible version of this figure.*

curriculum components—essentially saying, "We do all of this already." This is a misstep, as it will not lead toward additional growth or foster a team mentality that is focused on problem solving and the higher-order critical thinking you want to nurture in your social studies classrooms.

Identification of Students' Literacy Skills

To dive into disciplinary literacy, your team needs insight into your students' literacy skill strengths and weaknesses. Literacy-based inventories or assessments that might show up on a standardized test, like the grade-level exams states often issue or even college-preparatory exams like the ACT or a reading passage from the SAT, will help you identify where students are in terms of literacy skills so that you can determine the necessary strategies to help students navigate challenging text tasks.

Many schools using the response to intervention (RTI; Buffum, Mattos, & Malone, 2018) or multi-tiered system of supports (MTSS; National Center on Intensive Intervention, n.d.) frameworks already have literacy benchmark assessments in place, although the data are not always shared among all departments in the school. These assessments typically hold a wealth of information pertaining to student strengths and weaknesses for teachers of all content areas. If your school has an intervention specialist on staff, he or she would be a great colleague to approach to learn what data your school has already collected and what other resources might be available.

In the absence of such data, your team might work with a literacy specialist to design your own literacy-based inventories, employing your content area's authentic texts and assessing the literacy skills most pertinent to the specific field of study. Studying these initial student data will be a key component to your team time. Such data help your team identify which strategies scaffold student success and develop and utilize a variety of assessment types, including formative and summative.

Confidentiality and professionalism are necessary for a productive collaborative model. Getting to know your students better means looking at and sharing your student data with your team. When examining and collaborating around assessment data, it's important for teams to maintain a judgment-free zone, creating a safe space for team members to confront any issues the data reveal. During team meetings, it is critical that no one makes sweeping statements or generalizations regarding teaching practices based on raw data. When viewing group data in a team discussion on progress, remove names or class cohort information. Start the data discussion by noting this information may reveal things the team has already considered. For example, rarely does a nationally normed assessment show that a star student is one of the most struggling readers. Begin the data conversation with the idea that the data will often confirm what you already know about your students, but it may also shed more light on *why* students are struggling.

As your colleagues become more comfortable with data reviews, any fears about sharing data among team members will gradually drop away, and the team will become more open to collaboration that is focused on student learning. If you are a facilitator or literacy leader on the team, start with yourself; don't be afraid to show your own student data. Students have entire academic histories prior to meeting their current roster of teachers. There is no way that "one month in Mr. Williams's class" dictates all aspects of a nationally normed score.

Although it is critical that teams view data in a supportive manner, make sure that your team's analysis doesn't look like a list of excuses. It is very easy to fall into the pit of "things we can't control" when looking at less-than-optimistic numbers. Instead, use undesirable data outcomes to examine what your team *can* change; the discussion should focus on how team members move *all* students toward a goal, target, or outcome—and ultimately toward graduation, higher education or training, and a career.

Identification of Content-Literacy Connections

In our experience, social studies teachers understand the importance of literacy skills to a social studies curriculum. However, their views still usually reflect a traditional view of the concept of literacy. When we refer to *literacy*, we mean the act of engaging, knowing, and ultimately being able to navigate new understandings of known and unknown nuances associated with defined content. Yes, this sounds complex! But really, teaching literacy is as much about breaking down an idea, only to build it back up again by scaffolding and modeling a process—just like teaching a child how to tie a shoe. When modeling, teachers break down the processes for a content or process standard, step by step, demonstrating how to achieve the standard's goal. In literacy, modeling for students helps demystify the reading, writing, and thinking necessary for success for a task. We chose or designed the strategies we provide in chapters 3–6 to help model the reading and writing process for students, including how to make adaptations for both struggling and proficient students. All this said, words are literally everywhere, and many literacy texts are multimodal. Formulating an inference from a reading is similar to formulating an inference from what someone might say from a documentary or from a scientific demonstration. In comparison, literacy strategies we use to understand or to infer are often similar no matter what the modality or concept might be.

After your team has established a solid common baseline knowledge of the concept of literacy, focus on your power standards to identify which literacy tasks are the most innate to your area of study. Here is where you can identify your process standards and consequently outline the scaffolds that will support students' literacy skill development. Familiarizing yourself with the strategies that come later in this book will help you work through the process of finding your content-area literacy connections. Additionally, you will need to take a close look at your texts (solo, or as a team for common texts) to determine if they are appropriate for your readers and tasks. Gather all texts and assess your collection as a whole, using the tool in figure 1.2 as a guide.

---

1. Does content match your defined content power standards? If not, what is missing?

2. Are these texts at an appropriate text-complexity level for your readers?

3. Are your texts varied to address the different types of content associated with your discipline?

4. Once you have gathered your texts, what scaffolds will your students need to comprehend the content successfully?

5. What scaffolds will your students need in order to transfer knowledge to new contexts and applications?

Figure 1.2: Review tool to discern whether text tasks match complex disciplinary literacy demands.
*Visit **go.SolutionTree.com/literacy** for a free reproducible version of this figure.*

Having a firm grasp of how the readers in your classroom will navigate both text and the task connected to it is a critical component to scaffolding and modeling successful reading comprehension essential for content mastery and, more importantly, for skill and theoretical concept application.

Wrapping Up

If a fully supported and committed team dedicated to literacy sounds like a far-fetched dream, that's OK. All you need is one colleague to join you in your efforts to formulate collaboration that leads to positive changes! As your team collaborates, be sure to share its journey and findings with colleagues. Collaboration isn't always as formal as a designated team time—it often starts with an individual teacher simply sharing what he or she is working on and trying to accomplish with students. This work involves a lot of give-and-take from all team members. By conducting a thorough analysis of your curriculum standards, texts, and students' skills, your team will be well on its way to creating a common understanding of where students are. It will then be able to determine how to move students toward disciplinary literacy.

Effective teams work together diligently and value all contributions in their quest to help students succeed. Building a productive team is essential to working collaboratively toward teaching literacy skills rather than only the study of literature. To do this, collectively unpack the social studies reading, writing, and language standards to determine the essential skills (building blocks) inherent within them; set appropriate goals for students; and develop tasks that foster student growth.

Collaborative Considerations *for* Teams

- Who are colleagues you can approach to begin fostering collaborative teamwork?

- What social studies standards require literacy skills that lend themselves to action research for a team? (You will need to prioritize.)

- If you already have a social studies or literacy-based team, how is it organized? Is there a PLC culture that supports dialogue and all ideas that might lead to improved student outcomes?

- How can you use the resources in this chapter to develop your team commitments and your PLC culture?

CHAPTER 2

Foundational Literacy Triage

Throughout this book, we focus on how to establish a culture of literacy in the social studies classroom as well as throughout the school. All educators committed to learning must recognize the value of literacy, and all need to celebrate the results when an entire school establishes goals and creates an environment for students to feel safe when attempting and learning literacy strategies. However, this process takes time, and we often have teachers come to us who need immediate assistance. They have specific students, classes, projects, or assignments they are struggling with, and they need a strategy or way to help students *now*.

Always keep in mind the array of learners in your classroom. When instructing students in a mainstream classroom, there will be a range of student learning proficiencies. Further, students' reading levels may be even more diverse. Among these reading levels, you will have students with social-emotional needs, learning impairments, students with a 504 Plan to address a legal disability, and students who are only just learning English. All of these student groups are capable learners, but they all have different needs as unique as the learners themselves. As social studies teachers, your team must meet the needs and developing potential of *all* the students in your classrooms.

We have seen that teachers often begin to notice concerns related to literacy early in the school year and look for differentiated instructional ideas to meet student needs with the necessary supports. This chapter helps teams identify ways to *triage* literacy within social studies classrooms immediately and work with the variety of student needs and abilities found in your classrooms. We cover the basics of RTI practices, differentiated instruction, assessment of text complexity, and several fix-up strategies to assist social studies students when reading a text (Tovani, 2000).

We close by examining factors to consider when supporting struggling students and when enriching the learning of students who show proficiency.

Understanding Response to Intervention

RTI, also referred to as MTSS, is a three-tiered systematic process for ensuring the time and support students need in order to learn at high levels. Educational consultants Austin Buffum, Mike Mattos, and Janet Malone (2018) explain, "Tier 1 represents core instruction, Tier 2 represents supplemental interventions, and Tier 3 represents intensive student supports" (p. 2). Figure 2.1 illustrates the inverse pyramid Buffum and colleagues (2018) present for RTI practices.

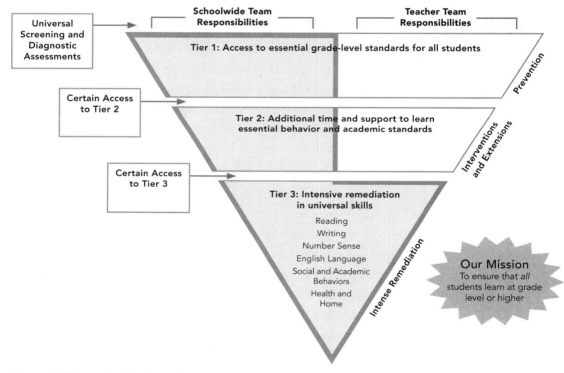

Source: Buffum et al., 2018, p. 18.

Figure 2.1: The RTI at Work pyramid.

Those familiar with more conventional RTI content may not be used to the inverted pyramid of RTI at Work. Of this inversion, Buffum et al. (2018) write:

> The traditional pyramid seems to focus a school's intervention system toward one point: special education. Subsequently, schools then view each tier as

a required step that they must try to document prior to placing students into traditional special education services. Tragically, this approach tends to become a self-fulfilling prophecy because the organization starts interventions with protocols designed to screen and document students for this potential outcome.

To challenge this detrimental view of the traditional pyramid, we intentionally inverted the RTI at Work pyramid, visually focusing a school's interventions on a single point—the individual student. (pp. 18–19)

Because a student's literacy skills are often linked to their learning in other academic areas, thinking about the role of literacy in RTI is an important topic to cover when students need support. We believe that teaching literacy is the responsibility of all teachers, and we think working with literacy interventions is a Tier 1 commitment—that is, all students should have access to grade- and course-level instruction related to literacy. As we noted earlier, teams often need more intensive literacy-based strategies to support students who might struggle or challenge readers who excel—this is where Tier 2 or Tier 3 interventions, remediation, or extensions might be necessary. Identifying students who will benefit from more intensive support in areas of literacy is an important consideration early in the school year—especially in social studies courses, where texts often require high-level comprehension that can challenge students.

Gathering data about students' reading and writing skills can be very helpful in learning more about the range of abilities in the social studies classroom. Data on nationally normed tests, such as MAP (www.nwea.org) and Star (www.renaissance .com), can provide social studies teachers with early insights into students' literacy skills. Likewise, teams should collaborate around creating common formative assessments that focus on measuring literacy-based skills important to the social studies curriculum they are teaching. Throughout this chapter, we outline a number of literacy-based concerns teams should give their attention as students are starting the school year, and we offer suggestions for responding if and when you and your teams notice other concerning patterns in student learning.

Differentiating Instruction

A good place to start toward meeting students' immediate, varied needs is to consider how you can differentiate instruction. *Differentiated instruction* is teaching that helps students with diverse academic needs and learning styles master the same challenging academic content (Center for Comprehensive School Reform and Improvement, 2007). It also provides students with interrelated activities that are based on student needs for the purpose of ensuring that all students come to a similar grasp of a skill or idea (Good, 2006). Differentiating does not mean providing separate, unrelated activities for each student. Instead, differentiation supports students with learning differences and helps them retain content and skills essential to the learning. We find differentiation of materials reduces the overall time students require to learn information, lessens the need for skill remediation, and allows students to demonstrate their learning in multiple ways.

There are four planning steps you can use when differentiating classroom work.

1. **Determine academic content or skills:** The first step in differentiating instruction is knowing and understanding the specific academic content you want to teach. This means knowing what learning targets (process standards) you want students to accomplish. What is the end result you want for your students?

2. **Gauge students' background knowledge:** You need to have a basic understanding of what the students currently know about the topic or content. At times, students may have no background knowledge on a topic. Consider issuing a benchmark assessment or survey that will help to determine students' readiness for new content.

3. **Select suitable instructional methods and materials:** Once teacher teams have an understanding of what students already know, instructional methods need to bridge what students currently know about the topic and what you want them to learn during the lessons. Knowing what knowledge gaps students have or what information they lack as well as where they are already proficient will help you plan to use specific methods and materials that are appropriate for instruction.

4. **Design ways to assess skill mastery:** After delivering instruction in multiple ways, the last step is to assess students' accumulation of new skills and knowledge for mastery. These formative assessments can take many different forms and be as simple as an exit slip or a quick response to a learning prompt. We cover assessing social studies students in chapter 7 (page 157).

When working with these four steps of differentiation, the process overlaps. Students all bring different knowledge to the content, work at different paces, and understand the material at different times. The overlap makes it possible to tailor the process to fit each student's needs at a particular time. In the following sections, we examine each of these steps in greater detail.

Determine Academic Content or Skills

As described earlier, PLCs focus collaborative teams on answering four critical questions. The first thing a team should ask is, "What do we want all students to learn?" (DuFour et al., 2016). In essence, what do we want them to know, understand, and be able to do? As a team of social studies teachers approaching these questions, determining the academic content or skills (or both) is the first step toward focusing decisions around curriculum, instruction, and assessment choices. In addition to determining power standards (essential standards) as we described in Standards Analysis and Goal Setting (page 24), we find that aligning literacy-based strategies to support these choices will help teams support students in applying literacy skills to social studies standards.

As you and your team work through this book's strategies, consider how certain strategies are better suited for the disciplinary content and skills you and your team are focused on teaching students. For instance, if your team is working on students' ability to synthesize complex information, it should work with strategies that help students make accurate and clear summaries. If the team wants students to draw connections between the past and the present, it should support students with literacy strategies that are geared toward compare and contrast and analysis.

Gauge Students' Background Knowledge

One challenge in teaching social studies is that it's common for students to have limited background knowledge of world history, their own country's history, government, religions, psychology, economics, geography, and sociology, making the teaching of new content in any of the particular content areas difficult. When students' background knowledge is limited, they struggle to make immediate connections or draw quick associations to what might be familiar to them, making it a challenge for them to add to their existing understanding or depth of knowledge.

When students have limited background knowledge in social studies topics, teams must recognize that they need to approach prereading strategies in ways that can draw connections between students' interests and their limited understanding of a topic that might be new to them. Teachers and teams can easily assess students' background knowledge using the prereading strategies in chapter 3 (page 45).

Although the use of these strategies presents an extra step for classroom instruction, they present a creative option to merge literacy and disciplinary learning.

Select Suitable Instructional Methods and Materials

Selecting suitable instructional methods and materials demands that teachers and teams ask a few different questions. For example, when considering literacy skills, selecting suitable materials means differentiating and aligning reading and writing tasks to the appropriate ability level of your students during their grades 6–12 experiences. Pay close attention to the reading level of the materials you select for students; they should be at or slightly above grade level. Students who require more support might need supplemental readings at an easier reading level that will scaffold them up to the grade- or course-level reading. Students who are more proficient readers might benefit from readings that extend their understanding of the social studies content.

One benefit of working on a team is that team members can collaborate to locate a variety of possible readings to help all students advance in their abilities. Likewise, drawing on the expertise of teammates can help inform the differentiated instructional methods each team member uses to help all students develop their potential in a variety of social studies topics.

In considering instructional choices, focus your team on discussions that support students in ways that advance their abilities to confidently approach simple texts and simple questions, and complex texts and complex questions, with greater confidence. In our work with social studies teachers, we find this approach to differentiated instruction helps all students, illustrating that there is no one-size-fits-all instructional strategy.

Design Ways to Assess Skill Mastery

Assessing skill mastery addresses the second critical question of a PLC: How will we know when they have learned it? (DuFour et al., 2016). In our work around literacy, we encourage teachers and teacher teams to focus on assessing skills in formative ways, often and over time. Although we specifically cover assessment strategies in detail in chapter 7 (page 157), many of the strategies we present throughout the book have applications in formative assessment. When teachers work with formative assessment practices, they can respond more immediately to the learning needs of students and support students quicker and more effectively.

Remember that through your team's commitment to teaching, assessments are meant to encourage collaborative conversations about the literacy of students in the social studies classroom. This collaboration helps to focus discussions on how team members can support all students and work to address the third and fourth critical questions of a PLC: What do we do if students haven't learned yet? and How do we extend learning for those who are already proficient? (DuFour et al., 2016). Well-designed assessments engage teams in developing and extending the potential of all students. As social studies readings tell a story and help students understand their world, it is important for the assessments to communicate information that helps teams understand what the students know and don't know.

Consider your most recent reading assignment. How can you differentiate the reading for students at multiple reading levels within your classroom setting?

thinking
BREAK

Assessing Text Complexity

An important factor to consider when seeking immediate assistance with instructing your students is making sure the reading level of the text is appropriate. This information may be available to you at your fingertips. Many times, the publishers of a book provide a Lexile measure or reading-level score, and if a resource doesn't, you can usually determine its level using search tools at the Lexile website (https://bit.ly/2PgXSqw). Equally important is knowing the Lexile level of your students. Having this knowledge allows you to appropriately select materials that challenge students without frustrating them—materials that are at or slightly above their Lexile level. You can determine individual student Lexile levels through standardized testing measures. (Visit https://bit.ly/2ZmS5EF to learn more about this.)

With knowledge of each resource's Lexile level and the Lexile level of each of your students, the challenge for teachers and teams is to provide a range of texts on a topic that fits the differentiated needs of students so that no text is too complex (students do not learn the material needed to bridge the gaps in their learning and literacy skills remain stagnant) or too basic (students do not advance their knowledge or literacy skills during their learning time).

Many online tools are available to help teams find text sets around a social studies topic with a range of different reading levels. Such resources allow teachers to support varied reading levels to aid students' understanding of a topic or to extend the complexity of understanding of students who have a stronger grasp of a topic.

Newsela and CommonLit, which we detail in the following sections, are two great resources that are immediate and easy to search on the internet.

Newsela

Newsela (https://newsela.com) is a large database of stories and articles about current events designed for classroom use. The site arranges articles by general themes, such as social studies, arts, health, science, law, money, entertainment, weather, and so on. They are student friendly and often provide different options on the same topic that can support students at many different reading levels. This allows all students to access the text at their level and still be reading, discussing, and receiving instruction on the same topic, making it an ideal resource for differentiated learning.

The website is available in a free version as well as a paid version for more specific readings and questions. It offers a variety of search options for texts based on your selected topic. You can also select the general grade or course level, text level, reading skills you want to focus on, and specific languages in which you would like the article to be available to students. In addition to searching for a specific topic, Newsela has text sets available. These sets offer an essential question, supporting questions, student instructions, and extension resources. (To learn more about essential questions, see Frontloading With Research, page 60.) Each text set includes multiple articles on a general topic. These text sets are specific to each content area, and there are separate social studies areas for instructors and students to view and select articles.

CommonLit

CommonLit (www.commonlit.org) is another free online tool for school use. It focuses mainly on grades 5–12 reading and writing within different content areas—to be sure, social studies teachers will find this a strong resource. Team members can pick passages by grade level, themes, text sets, and standards. CommonLit allows teachers to access texts of high interest that are aligned to standards. Students can annotate CommonLit articles, and many texts include questions or other activities. Users can search for texts categorized into a wide variety of themes (for example, the United States, justice, freedom and equality, resilience and success, social change, and revolution) or text sets (for example, U.S. colonies, the American Revolution, ancient civilizations, nature and conservation, space, and westward expansion). This online tool is great for finding and selecting texts at various levels and text sets around a common topic.

Using Fix-Up Strategies

Fix-up strategies—strategies to help students get unstuck during independent reading when they find a text confusing—are another way to assist social studies students when reading a text (Tovani, 2000). They provide a way to not only model but also teach good literacy habits and strategies to students by using the fix-up strategy in conjunction with a complex text. For example, when reading a World War II-era primary source, a history teacher might demonstrate a connection between it and a reading from a previous unit about World War I to demonstrate for students how to construct meaning and utilize and build on prior knowledge to overcome a difficult spot in the text. Cris Tovani (2000) writes that modeling in this way can support reluctant readers by providing more explicit instruction and reminders to use fix-up strategies, which include the following, phrased in student-friendly language.

- ▶ **Make a connection:** Find connections with what you are reading to other texts, experiences you have encountered, or the world around you.

- ▶ **Make a prediction:** Predict what will come next in a text or make a prediction about an upcoming reading based on knowledge gained from a previous text or class discussion.

- ▶ **Stop and think:** When something doesn't make sense, stop and think about what you've just read.

- ▶ **Ask yourself a question:** When something doesn't make sense, ask yourself a question about the text. Try to answer the question using what you know or new information you've found in the text.

- ▶ **Reflect in writing on what you've read:** Reflecting on your own thoughts about what you read can help to solidify connections to a reading and identify what might be new. It can also give you an opportunity to formulate an opinion. Reflect on what you might agree or disagree with, on what stands out in the reading and why, or on what makes you think differently about the topic and why.

- ▶ **Visualize:** Stop and create mental images of the concepts described in the text.

- ▶ **Use print conventions:** Examine headings and emphasized words to identify important ideas in the text.

▸ **Retell what you've read:** To help build the ability to summarize and synthesize, after you read something, stop and retell someone (or yourself) what you read in your own words. Try to capture the key point and important details in a way that makes clear sense to you, the reader.

▸ **Reread:** If something doesn't make sense, reread for clarity.

▸ **Notice patterns in text:** Consider whether reading is establishing a pattern. For instance, does the reading argue both sides of a debatable topic, and are there patterns to how the reading presents those viewpoints? Do patterns in the reading establish cause-and-effect relationships? Why or how does the pattern help guide the development of the reading?

▸ **Slow down or speed up:** Adjust your reading to slow down or speed up depending on your level of understanding.

Fix-up strategies are always a good reminder of how to read a text. Team members can create bookmarks for all students, listing these strategies as a quick reference to use when reading. A tangible bookmark in their hands when reading helps remind students to use literary strategies consistently. When they do so, these strategies become habits students can apply whenever they struggle with reading comprehension.

When using fix-up strategies, we recommend several different ways to make sure students are using them effectively.

▸ Help students to realize that during-reading strategies, like those we present in chapter 4 (page 81), are important *skills* to learn. Developing during-reading skills can help students with their comprehension and critical-thinking skills. When first teaching these strategies, help students to understand the value the strategy has for their learning.

▸ Model how to use the skill. Try not to assume that students will simply use the strategy. Take the time to model *how* they can use the strategy. For instance, to model how to read and ask questions or how to read and retell, a teacher might take a portion of the text and annotate it in front of the class to record questions and explain his or her thinking process. Often, students can practice these during-reading strategies with collaborative partners in class. With all of the social studies—specific strategies we present in the forthcoming chapters, modeling each of the habits is also good practice, and it provides students with examples of what they can do and what teachers expect of them.

▶ Encourage focusing students on using one or two during-reading strategies at a time. Consider the reading you are giving students, then align one or two strategies that will be most effective for that particular reading. Ask students to practice that strategy.

It's important to remember that you and your team can adapt all of these habits or strategies to meet the specific needs of your students in the immediate classroom setting. For instance, a teacher can adapt and use the retelling strategy for use in small groups or in partners, where everyone reads, and when one student retells the group what they read.

What additional questions or thoughts, at this point, should you consider when you approach your administration about creating a schoolwide literacy program that can specifically assist in the social studies classroom?

thinking
BREAK

Considerations When Students Struggle

Some students will likely continue to struggle and need repetition with these strategies even after you implement differentiation, find appropriate texts for their reading levels, and share fix-up strategies. As with implementing any new strategies or processes, these struggles may persist as students begin learning how to read and process a text in ways that help develop a more attentive level of comprehension and the ability to think *while* reading. This takes practice and commitment, which is why we think all teachers should work with students on these strategies for learning. Some additional questions to consider when you encounter students having a hard time accessing the text or using basic fix-up strategies include the following.

▶ **How might you help students recognize what level of text they are reading or able to access?** One way is to notice the vocabulary level of the text. If the student struggles with too many of the content-specific vocabulary words in the text, the reading might be too difficult for him or her, and the student might need support. Various source documents and social studies texts are often filled with new vocabulary words and concepts, so working on vocabulary development in social studies readings is often an important commitment to student learning.

▸ **How might you provide immediate help to students who have trouble reading or understanding specific social studies vocabulary or concepts?** There are many ways to improve both vocabulary development and understanding of concepts, such as by having students experience the word in differing forms and contexts. For example, you can provide two readings over the same topic, ensuring specific vocabulary appears in both readings but in different contexts. This helps the students to see, to read, and to hear the words in new and different ways, deepening understanding of the vocabulary. Likewise, students need to *use* new words they are learning. Consider building *word walls* around your classroom—charts of paper or huddle boards that list important vocabulary words and social studies concepts students should be learning to help support their comprehension and understanding of social studies. Refer to the word walls and ask students to use the words or practice techniques regularly during class time and within their group work. Frontloading, which we introduce in chapter 3 (page 45), is also a useful prereading strategy to introduce new vocabulary. Remind students that to think like a social studies expert, they need to talk and use vocabulary like a social studies expert.

▸ **How might you help students stay focused during reading and use the fix-up strategy bookmark you provide?** Have students use the bookmark as a reminder or reference for what students should be thinking about while they read. When establishing a purpose to read, ask students to refer to the bookmark as a way to keep their mind focused on that purpose or on the key questions they should be considering.

Although these strategies are helpful, some students will continue to struggle with the demands of literacy, especially as social studies texts and concepts increase in difficulty. When working with students who might continue to struggle, it is important to intervene and anticipate difficulties they might confront, including determining if students lack course- or grade-level skills that might require higher-tier intervention.

As an individual teacher and as a team, work to prepare supports that will help every reader to succeed with comprehending and understanding the literacy tasks expected of him or her. In setting expectations for reading that are directed and purposeful, teachers can help students achieve at a rate that is individualized through appropriate challenges. A reader who is still developing might need to focus on simpler questions during a first reading and then focus on a more complex question during a second reading. Or, a developing reader might need clear direction on three important points to identify so he or she can then find them in a text. In working with varying reading levels, team collaboration can help with generating new and fresh approaches to the assigned literacy tasks.

Considerations When Students Are Proficient

Just as it is important to support the developing readers in your social studies classrooms, collaborative teams want to develop the potential and critical-thinking skills of students who demonstrate proficiency and mastery. In social studies, consider collaborating with your team to create complex questions and tasks that will challenge students to extend their critical thinking and capacity. Your team might have other readings that advance students' disciplinary knowledge, but you might also ask the students to process their thinking about a text in more complex ways. For instance, ask students to respond to higher-order-thinking questions that are focused on building connections or associations between social studies concepts.

As a team, remember that your collaborative work should address the developing potentials of all students. Often, by focusing on more advanced ways of thinking, teachers and teams will uncover creative and interesting insights into revising a curriculum to be more engaging, too. These extension opportunities for proficient students can help lead the team in overall curriculum design and writing or rewriting curriculum when needed.

Wrapping Up

By establishing a triage approach to differentiating instruction and finding texts at students' varied reading levels, your team establishes a powerful way to open up new opportunities for the reading process. If you match these with fix-up strategies and share a bookmark listing these strategies, you can equip students for a more successful reading process and help them gain the confidence to continue reading their social studies texts. Modeling the fix-up strategies is key to showing students how to use them. Continuing to provide opportunities to practice and repeat strategies will help students learn the strategies as ongoing habits. In this way, the strategies become more automatic to their own maturing reading process.

Collaborative Considerations *for* Teams

- What fix-up strategies seem most appropriate for a social studies text?

- How might you adapt the texts you use for different reading and ability levels?

- What are some good social studies texts your team might use that are appropriately targeted challenges for your classes?

CHAPTER 3

Prereading Strategies

As we wrote in chapter 1 (page 17), one major aspect of the journey to teaching students to think like social studies experts began for us when we asked our collaborative team how to find the appropriate balance between content and skills. In our experiences working with history teams, many teachers had moved away from traditional textbooks. Instead, teachers provided students with much of the course content via slideshows and lectures. This was, in part, because students were struggling with the textbook, but it was also to allow for more instruction around primary-source reading. Many of the teachers on the team expressed frustration that their students had come to depend on the teacher as purveyors of content as opposed to having the capability to garner essential knowledge from the textbook.

We collaborated with the U.S. history team first, and we set out to come up with several prereading strategies that would help students of various reading levels successfully navigate the variety of texts they would encounter in class. Focusing on prereading made a lot of sense because we know from experience effective readers of all content areas must make a variety of prereading moves to effectively comprehend new material.

This chapter explains the need for prereading in social studies, including schema theory. After discussing the rationale for social studies prereading and the need for collaboration when combining social studies and literacy instruction, we examine the four types of social studies prereading strategies. We also highlight the C3 Framework for Social Studies and how it aligns with the CCSS ELA to highlight embedded prereading literacy skills in the standards, and we offer concrete strategies we developed through team collaborations with our social studies teachers. The chapter concludes with considerations for addressing students who continue to struggle and those who show proficiency.

The Importance of Prereading

One thing that students, and sometimes even teachers, forget is that prereading is an essential part of the reading process (Urquhart & Frazee, 2012). In prereading, strategic readers set the stage and prepare their minds to take in and negotiate new information, and this is something students can approach in several ways. On one level, it can be as simple as surveying and previewing a text to gain an understanding of the content, helping the reader to understand the scope of the text and its depth. In other ways, prereading strategies might be more creative to engage readers and spark their interests. Further, prereading strategies may be more pointed and specific, clearly establishing a purpose to the reading task or identifying what specifically students need to learn from the reading. In grades 6–12, prereading strategies are important for beginning readers as well as more advanced readers because they help students frame their commitment to learn from the text, be it simple or complex. Prereading strategies prepare students for the reading experience and what they will encounter once they are engaged in the text.

From your teacher preparation classes, you might remember the concept of *schema*, the organizational network in the brain that takes in, stores, and connects information to other ideas (Rumelhart, 1980). It is not unreasonable for teachers to teach their students about schema and make the concept understandable and relevant to them. Sometimes in our work as literacy coaches, we refer to schema as *brain glue*: if students want information to stick, they have to get in the habit of activating their brain glue.

To reinforce this concept, you can regularly model and share your own reading processes with students throughout the school year—for example, when talking about historical or other nonfiction texts, when discussing new or relevant articles, and so on. When you regularly model the concepts, then you can expect prereading activities to activate brain glue as students begin to build these reading behaviors and become more proficient readers. When modeling, perform for students exactly what they will be doing when they work independently. Narrate the thinking process, including challenges they might face and how to overcome them.

Prereading strategies also help to activate prior knowledge and experience to help the reader make connections to new ideas or new information, and they work to prepare the reader in ways that help him or her focus with greater intention and deepen comprehension and understanding (Ferlazzo & Sypnieski, 2018).

Collaboration to Combine Literacy and Social Studies Expertise

Collaboration is a central component of PLC culture and the development of team structures. When experts collaborate, they learn from one another and consider innovative ways to approach student learning. As experts in literacy, we've found that many teachers from all subject areas have some knowledge of basic and intermediary literacy skills, like the ideas we discussed in chapter 1 (page 17), but we also find many teachers need support to adapt and develop literacy strategies that are specifically tailored to their subject area and the difficulty level of the disciplinary texts. For instance, in teaching our students to be more effective readers of social studies content, we asked the U.S. history collaborative team to explore and retrace its own moves as proficient readers of history so we could adapt their own basic strategies and skills to fit students' needs and mentor them to read like disciplinary experts. We asked the U.S. history teacher team to pose and discuss the following questions among team members.

▸ "How would we read this text?"

▸ "What would we have to know to understand the reading?"

▸ "How would we annotate or organize the information?"

We wanted to know more about how the team approached the reading of social studies texts, how they learned from reading about social studies, and what they did after they finished reading a text. From them, we gathered several insights about how they approached the reading process in ways that could direct our collaboration and decision making about prereading, during-reading, and postreading strategies. Together, the teachers on the team learned a valuable lesson about their need to understand and monitor their habits as readers if they were to teach students to successfully read social studies texts. As professionals, teachers may take for granted the prior knowledge and skills they bring to a text because reading history and other social studies nonfiction texts has become second nature. Once the U.S. history team realized this aspect of reading like a historian, members felt empowered to collaborate around the ways they would help students to navigate the reading.

The work our U.S. history team initially wanted to focus on was centered around readings common to the ACT test, including sample readings and questions to which the team had access. The teachers were concerned that students were struggling to be successful with these high-stakes social studies reading passages. Note that, in the time since we began our work, many classrooms we've worked with have moved their focus from the ACT to the SAT; however, the core principles of reading dense social studies texts that we learned as a team remain relevant to both.

Initially, we sat together for a long time and discussed how we would navigate these types of texts, starting with an adapted piece from volume two of Blanche Wiesen Cook's (1999) biography of Eleanor Roosevelt. It was amazing to hear the variety of ways that experts would go about the task of reading such passages, but the thing that all teachers had in common was that they would engage in some sort of prereading.

Before jumping into the passage about Roosevelt, some teachers would make *predictions* about the content of the passage based on their *prior knowledge* regarding the subject. Of course, as a group of historians, they knew quite a bit about the former First Lady; however, they did not know exactly which aspects of Roosevelt's life that this particular passage would focus on. Those of us who had less knowledge about Roosevelt sought out short online videos or summaries to help us *preview* what the text might be about. An online search engine provided us with everything from a simple Wikipedia entry to a four-minute video to feature-length biographical films. Perusing the shorter pieces provided us with plenty of background knowledge to set us up for success with the Roosevelt piece. Finally, another popular strategy the teachers used is to look ahead to the ACT questions at the end of the passage. By knowing the questions they sought to answer, they developed a focused *purpose* for the reading. Through this process, we began to see the different needs and strategies the team could use to mentor students in prereading.

The collaboration we highlight here can support any social studies team in finding consensus about the most impactful prereading strategies for a curriculum. For example, using the process we describe, our social studies team could subsequently tackle other similarly dense reading passages while mentoring students to read like historians, equipping them with various prereading strategies students could use as they progressed as readers.

The Four Ps of Social Studies Prereading

As our collaborative team example in the previous section highlights, when approaching literacy strategies, four prereading strategies are particularly important to consider: (1) *previewing*—scanning a text to gain an overview, (2) *predicting*—considering what the reading will say and what information it might provide, (3) *prior knowledge*—identifying what the reader might already know about a topic, and (4) *purpose*—establishing a focused reason for reading. As you may have already observed, these four skills are not just for prereading: they permeate the reading process. Skilled readers constantly predict and connect to prior knowledge; less frequently, they also preview and review the purpose during the reading. The important thing to remember is that if readers are to navigate texts successfully, they must learn and use prereading skills in a variety of ways to ready themselves to enter the process of reading more successfully. Here are a few points to consider.

▸ **Model how to use prereading strategies with students:** Modeling aloud can help students understand how to use prereading strategies intentionally when approaching simple texts or more complex readings. Again, this means taking a few minutes to demonstrate the task for students, taking the time to explain the thinking process.

▸ **Focus on one or two prereading strategies with students:** While the reading process should be deliberate, no one wants it to become cumbersome, so teach students tactics for determining which prereading strategy will best fit a particular task. The goal is to use prereading strategies flexibly and effectively.

▸ **Prereading strategies should become habits that don't take excessive amounts of time:** Although it takes practice to achieve proficiency, prereading strategies are designed to help the reader frame a purpose for reading and a motivation to read. The real work of reading to learn occurs during the next step in the reading process.

Students should learn to preread in all content areas, including history, economics, psychology, sociology, and all other courses in the social studies curriculum (Urquhart & Frazee, 2012). The following is a basic guide to the four Ps of prereading that teachers can share with students.

▸ **Preview:** Students should scan the reading for various purposes. They could review subheadings, examine images, look for keywords, read captions, or identify specific disciplinary writing techniques they might need to understand. A good preview gives readers an idea of the content they are about to tackle. It may encourage readers to think about what they already know about the topic, it may lead readers to consider what the text is about, and it may help readers understand why they are reading the text and what the author hopes to accomplish. Recall the earlier discussion of the ACT passage on Eleanor Roosevelt.

Like history experts, your team would want students to preview and ask themselves what they already know about the topic and what they might need to do to prepare themselves to read the passage.

▸ **Predict:** Readers should constantly think about what they might encounter and learn as they read or view a piece. Anticipating what is to come next in a text is a good way to keep readers engaged and allow strategic readers to confirm or deny their thinking. This can lead to better comprehension and retention. This comes naturally to students when watching television and movies for pleasure because they are accustomed to anticipating how the story will play out. For some reason, many students do not predict the same way when reading a nonfiction text, but they *can* learn. For example, before reading about Eleanor Roosevelt, a teacher might ask students to predict the ideal qualities a First Lady might possess and then compare and contrast their predictions with what they learn about Roosevelt during the reading.

▸ **Prior knowledge:** David E. Rumelhart (1980) posits that all of our generic knowledge is embedded in our schema. More recently, experts argue that by accessing prior knowledge, students will be more equipped to tackle complex problems (Goodwin, 2017). By actively recalling this knowledge, readers are prepared to build on the knowledge they already have and store that new knowledge for later use. By recalling what they know already, students are better prepared to attack a more complicated reading and connect it to developing schemas. Also worth noting is that if students have rich background knowledge, they are generally better prepared to handle more complex material to add to that prior knowledge. If students have little prior knowledge, be mindful of the difficulty level of the reading so they are not quickly overwhelmed. Continuing with our previous example, most secondary students have at least some familiarity with President Franklin Delano Roosevelt, but many may not know about the First Lady. However, secondary students likely know some things about *the role* of a First Lady. By thinking about what they know about Franklin Roosevelt and the role of the First Lady, students activate some schema with which to connect the knowledge they will accumulate from the reading.

▸ **Purpose:** There are two aspects of purpose that are important for readers to understand. First, readers should have a clear purpose for *why* they are reading something. Based on our personal experiences, if a student's reason for reading is "because I have to," he or she is probably not going to get a lot out of the reading. However, if a reader can clearly articulate the purpose for reading, odds are he or she will be more fully engaged and benefit from the experience more deeply. Second, readers more fully engage with text when they can *predict* and *anticipate* the author's purpose for writing. Sometimes the author's purpose is obvious; for example, knowing that a primary source is a political cartoon published in a newspaper from a historical period is important for understanding the author wants to convey an opinion about a current event. Other times, a reader has to work harder to find the author's purpose through identifying a claim or thesis in the text. Explaining how to find an author's claim or message is a major component of the C3 Framework for Social Studies and the CCSS ELA skills because these skills provide students with a purpose that focuses their efforts when reading a text.

Think back to the last piece you read in your classroom. As a social studies teacher and expert, what did your team do to prepare for instruction (preread) before having students engage in a reading? If you cannot remember, think about a piece of reading your team plans to use in classrooms. What would each team member think about before reading it?

○ Do you notice team members gravitating to any of the four Ps? Why or why not?

○ What is most important for a historian, economist, psychologist, sociologist, anthropologist, or civics expert to think about before reading?

○ Does your social studies team collaborate around instruction that supports students before reading rigorous texts? What could you do to enhance prereading instruction?

thinking
BREAK

C3 Framework for Social Studies Connections

While there are fewer standards related directly to prereading than for during reading or postreading in the CCSS, research shows that effective prereading skills are a precursor to competent literacy because they prepare relevant schema to connect and engage with the text (Pardede, 2017). The CCSS College and Career Readiness (CCR) Anchor Standards for Reading, authored by the National Governors Association Center for Best Practices (NGA) and the Council of Chief State School Officers (CCSSO; 2010), are as follows.

- Read closely to determine what the text says explicitly and to make logical inferences from it; cite specific textual evidence when writing or speaking to support conclusions drawn from the text. (R.CCR.1)

- Determine central ideas or themes of a text and analyze their development; summarize the key supporting details and ideas. (R.CCR.2)

- Analyze how and why individuals, events, or ideas develop and interact over the course of a text. (R.CCR.3)

- Interpret words and phrases as they are used in a text, including determining technical, connotative, and figurative meanings, and analyze how specific word choices shape meaning or tone. (R.CCR.4)

- Analyze the structure of texts, including how specific sentences, paragraphs, and larger portions of the text (e.g., a section, chapter, scene, or stanza) relate to each other and the whole. (R.CCR.5)

- Assess how point of view or purpose shapes the content and style of a text. (R.CCR.6)

- Integrate and evaluate content presented in diverse media and formats, including visually and quantitatively, as well as in words. (R.CCR.7)

- Delineate and evaluate the argument and specific claims in a text, including the validity of the reasoning as well as the relevance and sufficiency of the evidence. (R.CCR.8)

- Analyze how two or more texts address similar themes or topics in order to build knowledge or to compare the approaches the authors take. (R.CCR.9)

- Read and comprehend complex literary and informational texts independently and proficiently. (R.CCR.10)

Compare these to the C3 Framework for Social Studies, which a broad committee of social studies experts created to provide states with supplemental guidelines to help students prepare for the challenges of college and career. Development of this framework began in 2010 and resulted in a set of state standards that challenges students to "know, analyze, explain, and argue concepts" in a social world

(National Council for the Social Studies, 2017). The C3 Framework for Social Studies (National Council for the Social Studies, 2017) represents an *inquiry arc*, "a set of interlocking and mutually reinforcing ideas" (p. 17) and features four dimensions of informed inquiry.

1. Developing questions and planning inquiries

2. Applying disciplinary concepts and tools

3. Evaluating sources and using evidence

4. Communicating conclusions and taking informed action

Since the National Council for the Social Studies (2017) wrote its framework after the CCSS ELA, which was published in June 2010, the developers recognized the responsibility of all content areas to develop literacy. To that end, each of the four dimensions in the C3 Framework for Social Studies relates to and extends appropriate CCSS ELA skills. Therefore, in the following tables, we provide some guidance to potential connections between the prereading strategies that we present later in this chapter and both the prereading skills in the CCSS ELA (table 3.1) and the four dimensions in the C3 Framework for Social Studies (table 3.2, page 54). Checkmarks in these tables indicate alignment between a specific strategy and a specific standard. While each strategy may not directly or initially connect to the standards checked, this provides some guidelines for teachers who can use their professional knowledge to adapt and tweak a strategy to hit the targeted standard.

Table 3.1: CCSS Prereading Strategy Connections

Prereading Strategy	R.CCR.1	R.CCR.2	R.CCR.3	R.CCR.4	R.CCR.5	R.CCR.6	R.CCR.7	R.CCR.8	R.CCR.9	R.CCR.10
Predicting via an Anticipation Guide	✓	✓	✓		✓	✓		✓		✓
Frontloading With Quick Writes		✓	✓						✓	✓
Frontloading With Research		✓	✓				✓			✓
Wide Reading and LibGuides	✓	✓	✓	✓	✓	✓	✓	✓	✓	✓

continued ⟶

Prereading Strategy	R.CCR.1	R.CCR.2	R.CCR.3	R.CCR.4	R.CCR.5	R.CCR.6	R.CCR.7	R.CCR.8	R.CCR.9	R.CCR.10
Frontloading Vocabulary With the Frayer Model	✓	✓		✓						✓
Visuals		✓				✓	✓		✓	
Inquiry Scenarios		✓	✓			✓	✓		✓	
Beliefs and Opinions Survey	✓	✓	✓				✓			
Turning Titles and Headings Into Questions	✓	✓		✓						✓
Possible Sentences	✓	✓	✓	✓						✓
Picture Books	✓	✓							✓	✓

Source for standards: NGA & CCSSO, 2010.

Table 3.2: C3 Framework for Social Studies Prereading Strategy Connections

Prereading Strategy	Dimension 1: Developing questions and planning inquiries	Dimension 2: Applying disciplinary concepts and tools	Dimension 3: Evaluating sources and using evidence	Dimension 4: Communicating conclusions and taking informed action
Predicting via an Anticipation Guide	✓	✓		
Frontloading With Quick Writes		✓		
Frontloading With Research	✓	✓	✓	
Wide Reading and LibGuides	✓		✓	
Frontloading Vocabulary With the Frayer Model		✓		

Visuals		✓	✓	
Inquiry Scenarios	✓			
Beliefs and Opinions Survey	✓			
Turning Titles and Headings Into Questions			✓	
Possible Sentences			✓	
Picture Books	✓		✓	

Source for standards: National Council for the Social Studies, 2017.

Strategies for Supporting Students in Prereading

In our practice as literacy coaches working with disciplinary teachers, we examined past practices teachers have used to introduce reading in the classroom. We examined how teachers might personally use prereading strategies when they read. And we explored other prereading options that might help to support their work with students in new, more innovative ways. For many social studies teachers, this examination is a powerful moment, because it's when they recognize the difference between what they do in the classroom, what they do personally as readers, and what they *could* ask students to do when reading for class.

Consider this: you know how you read, and you're an expert at it. But it's easy to forget that you also have to teach students how experts read. Therefore, one of the most powerful adjustments you can make as a social studies teacher is to begin learning about and creating prereading activities that guide students to become strategic prereaders. An array of different teachers and teacher teams created the following strategies and lessons for classroom use. Each of them combines the knowledge and experience of both literacy experts and content-area teachers. Each strategy also comes with an explanation and provides adaptations that allow for differentiation for students who might struggle, such as students who qualify for special education or students who are learning English, as well as for students who show high-level proficiency. We hope the strategies spark ideas for you and your team to use and adapt for your own classrooms.

Predicting via an Anticipation Guide

By making predictions based on prior knowledge, students can set a clear purpose for the reading that follows. Anticipation guides offer an approach to prereading that helps students focus on a reading's purpose, which continues into the during-reading and postreading phases. An anticipation guide asks students to read a series of assertions and make predictions about their accuracy. In the social studies classroom, anticipation guides support students in thinking critically about social studies concepts, thoughts, or ideas before reading about them, and students have the opportunity to see how accurate their predictions were based on the reading that follows. For example, an economics teacher may list several potential causes for a market's variations in supply and demand, and students must use prior knowledge and critical thinking to anticipate which of the statements seem true and which do not. Later, while reading, students can verify or correct their thinking. It also means they can focus their reading around evidence and key details that support or refute their predictions, allowing them to take more ownership of the reading.

How to Use

Start by providing students with a list of statements, asserted as facts. Figure 3.1, which organizes statements in a central column, presents a similar example based on the French Revolution. (See page 178 for the reproducible "Prereading Anticipation Guide.") Relying on prior knowledge and existing literacies, students use the leftmost column to indicate if they agree or disagree with the assertions, essentially making predictions. Providing students with items that are debatable or controversial often leads to the most exciting conversations as they fill out the anticipation guide or during a discussion following the prereading activity. During the reading process, students record evidence that either supports or refutes their prediction. After reading, students use the rightmost column to confirm or revise their initial assessment.

Adaptations

When adapting this anticipation guide for students who are struggling with the literacy aspect of the strategy, color-coding the Agree and Disagree columns can help them quickly identify the columns for indicating agreement or disagreement. Also, it may help students to complete this exercise orally for the first couple of items. Students articulating their thoughts out loud can help them better understand confusing ideas or concepts and clarify different thought processes.

Name: Pat Stevens

Prereading Anticipation Guide for the French Revolution

Before reading: Look at each statement carefully. Put a check in the appropriate column in Before Reading to indicate whether you agree or disagree with each statement.

During reading: In the center column, write the evidence from the reading that supports or contradicts the statement. Include the page number where you found your evidence.

After reading: Reread the statement and the evidence that supports or contradicts it. Put a check in the appropriate column under After Reading to indicate whether you now agree or disagree.

Before Reading		Statement and Evidence	Page Number	After Reading	
Yes	No			Yes	No
✓		1. People revolt against their government because they are fed up with their country's leadership. **Evidence:** Poor economic conditions and heavy taxes upset the majority who had a minority representation in government.	2	✓	
	✓	2. Revolutions bring positive change for ordinary people. **Evidence:** The Declaration of the Rights of Man ensured more equal rights, freedom of speech, and so on.	4	✓	
	✓	3. It is never acceptable to revolt against one's government. **Evidence:** The French Revolution demonstrated the power of the people to make a difference and set an example for future nations to follow.	4	✓	
✓		4. The French Revolution was successful in achieving its goals. **Evidence:** The revolution failed to meet all of its goals.	5		✓

Figure 3.1: Prereading anticipation guide.

For students learning English, it may be helpful to focus on vocabulary in the Statement column. Teachers can anticipate important or challenging words that are essential to understanding and include a balance of accurate and inaccurate statements. This helps students use context clues in an unfamiliar language to determine the meaning of words in the passage.

Encourage students who consistently and accurately provide evidence of proficiency with prereading skills to identify their own thematic statements based on their previewing of a reading assignment. The teacher can leave a few blank rows for proficient students to set their own purpose for reading that falls in line with the focus provided for the entire class. For example, in the sample in figure 3.1 (page 57), a proficient student might anticipate that *only minority populations can revolt*. Then, the student can look for evidence that supports or refutes this position.

Frontloading With Quick Writes

Giving students several minutes to make and write down predictions about a new reading allows a group with some prior knowledge to tap into what they already know. This makes frontloaded quick writes an effective way to support students as they set a purpose for the reading that will follow. Because students bring varying degrees of prior knowledge to a text, this strategy allows them to assess what they already know about a topic before taking on new material, and more important, it helps them sort and categorize that information as part of a frontloading activity. This can be an important strategy for social studies courses in that many students will remember different aspects of the history they have studied in the past. Since the strategy is not dependent on one correct answer, it allows all students to access their own knowledge.

How to Use

Give students a short quick-write prompt in the form of a question or a sentence starter. This prompt should relate to the concepts they are about to learn, and it can include vocabulary they will learn with the content or larger concepts about the reading. Teachers can narrow the focus or keep it as broad as they prefer. The purpose is to get students thinking and showing their thinking about what they may know, thereby establishing a sense of purpose around the topic.

For this activity, students have only a short amount of time to respond, perhaps five to ten minutes. It should be just enough time for them to get what they know onto paper as quickly as possible. Whatever the duration, encourage students to use the full time. At the end of the writing time, students can share out to the entire class what they believe they know, or they can get into small groups to share their writing with peers. Figure 3.2 shows multiple examples of prompts suitable for quick-write exercises in a U.S. history class. (See page 179 for a reproducible version of this figure.)

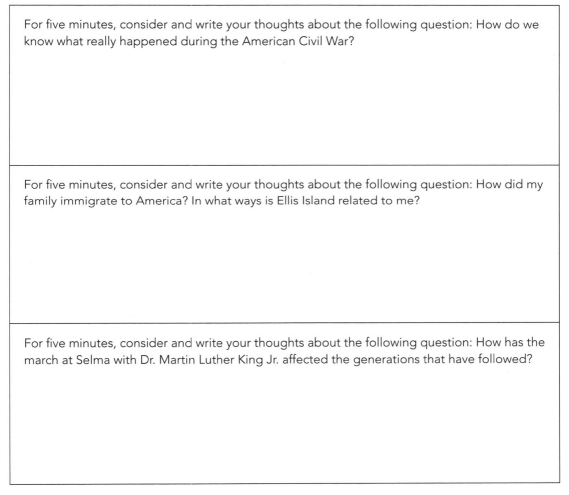

For five minutes, consider and write your thoughts about the following question: How do we know what really happened during the American Civil War?
For five minutes, consider and write your thoughts about the following question: How did my family immigrate to America? In what ways is Ellis Island related to me?
For five minutes, consider and write your thoughts about the following question: How has the march at Selma with Dr. Martin Luther King Jr. affected the generations that have followed?

Figure 3.2: Quick-write prompts.

Adaptations

Because student choice is empowering and engaging, we consider it best practice to provide struggling students with options by offering multiple quick-write prompts or sentence starters. This helps students create a sense of ownership over their writing and appeals to varying degrees of prior knowledge. Students in special education may need more structure in their written responses. Teachers can provide a sentence starter, but they should also add more sentence stems to help students expand on their thinking. Students will still consider the ideas addressed in the quick write, but they also receive additional writing support.

Additionally, finding a place to start can be so frustrating for students who struggle that they may not be able to write at all. For this reason, students learning English, in particular, may need teacher-provided information about vocabulary words or historical concepts that they will read or see in the text. Providing students with specific words in the prompts to use within their answers also gives them a starting place for their writing.

Even proficient readers will benefit from a simple quick-write activity because they benefit from the experience of warming up their minds to new information just as much as a struggling reader.

Frontloading With Research

Frontloading with research is an effective strategy to help social studies students explore inquiry questions for a lesson or unit of study without the need to rely on existing background knowledge. Through this strategy, students receive the opportunity to use a variety of resources to expand and build a new base of prior knowledge about a topic before engaging with the primary reading. It's also a perfect opportunity to instill in students valuable critical-thinking skills. Although 21st century students are digital natives and are generally very skilled at tapping into information, teachers still have an obligation to teach them *how* to sift through and evaluate that information to ensure its validity and reliability. Taking a portion of a lesson to model valuable research skills as a prereading activity, such as finding and evaluating the worth of an article, helps students understand the context of a particular social studies concept or historical moment before reading a primary document written about it. Consider using this strategy with the lateral reading during-reading strategy in chapter 4 (page 81).

How to Use

Provide students with an *essential question* about a forthcoming reading task—when a question is essential, it tasks students with learning and thinking about a topic beyond a surface level (Watanabe-Crockett, 2019)—and gives them time to explore the question using a variety of resources. These resources could come from materials you provide, such as other books, articles, music, or films, or they could come from students exploring online resources, such as articles, website searches, and so on. Be sure to use discretion when determining how ready students are to seek their own research resources. For example, you might offer the following prompt: *Who were the winners and who were the losers in the Arab-Israeli War of 1948? Read the article "Why Wars No Longer End With Winners and Losers" (Paul, 2010) and summarize the most important and compelling ideas that relate to this event.*

First, explain how the article is a valuable, reliable resource. Then provide students with an opportunity to summarize their findings and then share those findings with peers or use them as a reference during their primary reading activity. In doing so, students build new prior knowledge that enables them to set a purpose for the reading that is to follow.

Adaptations

Students need to understand what the question is asking before they can evaluate sources. Students who are struggling with the literacy aspect of the strategy may be able to build on their prior knowledge by working with the teacher or an aide to complete the task orally. Doing so can also give instructors a better understanding of where students are in their learning. To begin this process, state the essential question orally and explain what the question is asking. Second, clarify what some of the terms in the questions are. For the previous example, some students may not understand why Israelis and Arabs were fighting in the first place. Then, read a teacher-provided source out loud to students, conducting a think-aloud activity as you read the article. Have them list compelling main ideas or what they felt was interesting from the source. Once the essential question and research process are clear, students are better equipped to evaluate whether a source is valid and reliable.

Students learning English may benefit from a more limited menu of source material to ensure that the vocabulary and content are appropriate for their reading abilities. Although all students need access to full course- or grade-level instruction (Buffum et al., 2018), teachers may need to frontload instruction around vocabulary for more challenging texts. Online resources, such as Newsela (www .newsela.com), may also be helpful when adjusting the Lexile levels of scaffolded reading materials.

There is a good chance that the strongest readers have ample prior knowledge already, so encourage them to explore databases or the internet to find their own source material while other students search from a teacher-provided menu.

Wide Reading and LibGuides

Wide reading is a strategy that provides students more time to read and offers the opportunity to read a variety of texts around a specific topic. Education expert

Robert Marzano (2017) advocates wide reading as a major strategy for building academic background knowledge for students lacking experiences with important topics in the curriculum. Offering many different texts (books, videos, websites, pictures, and so on) provides access points for students to build knowledge before engaging in new learning or academic reading for social studies. More specifically, wide reading in the social studies classroom establishes an inquiry mindset toward learning—an academic curiosity to learn—that can create guiding questions or provide bridges into the specific learning targets and standards.

This strategy is very important for students with low prior knowledge or who have limited access to texts, technology, or media because it provides them with more resources, texts, and information to engage with. Teachers can pair wide reading with LibGuides, which are easy-to-use subject guides librarians create that contain many different texts around a specific topic or area of study. LibGuides may include a combination of websites, articles, journals, pictures, or any additional sources at various reading levels.

How to Use

Social studies teachers can implement wide reading in the classroom in a variety of ways. For example, teachers could create a library within their classrooms with a variety of fiction and nonfiction books and other reading materials for students to access. These texts should offer many different reading levels for students to choose from. Teachers should consider each student's reading ability when guiding him or her toward specific texts. All students should be able to read and understand the material they read without support. (See chapter 2, page 31, for more information on the usefulness of Lexile levels.) Students also need to be able to access online materials, whether through specific programs (such as Newsela) that offer a variety of resources at different reading levels or simply access to search engines.

Bringing your students to the school library, media center, or information learning center is another powerful way to show them how to access these multiple resources. Librarians are often-overlooked collaborative partners; they have access to a wide variety of materials that many teachers may be unfamiliar with. Additionally, technology and websites are always evolving. We have mentioned a few different resources throughout this book, but there will inevitably be new innovations and resources for students to access that emerge every year. Take advantage of your school librarians, and stay abreast of what can work best for your students.

Teachers can also ask librarians to create LibGuides for students to access information on a specific topic or area of study. For example, visit https://libguides

.d125.org/greatdepression to see a LibGuide about the Great Depression that is designed to scaffold learning for students learning English. It includes resources about the period, including links to images, videos, databases, vocabulary, and so on. Regardless of the method, during this prereading exercise, students need to see their teacher engaging in wide reading activities. With teachers setting the example by modeling wide reading and providing a variety of texts, students will be more willing to engage with different kinds of texts. In this case, modeling might involve previewing with students the various sources on the LibGuide, or it might involve showing the class how they can develop background knowledge by exploring the content on the site.

Adaptations

The key to wide reading being successful for struggling students, particularly students in special education or students learning English, is having various reading levels for them to choose from. Giving growing readers the opportunity to read material at their reading level, around the same concept or content as students working at grade level, can build a bridge to these students processing full grade-level texts. It also allows for small-group collaboration and conversation that further enhances learning. For this reason, we encourage using a range of texts that will appropriately challenge all students. It is possible to label each source with a Lexile score, so students can find a source that is appropriate for their instructional reading level. This differentiation allows all students—including those who show high-level proficiency—to come together and share their newfound knowledge. Share your enthusiasm for what you read (it's infectious!), and students will be more likely to develop their own enthusiasm for reading and learning.

Frontloading Vocabulary With the Frayer Model

When we meet with social studies teachers, they often tell us that students struggle with reading social studies and historical texts because the vocabulary is too challenging. Students may struggle to understand social studies concepts or definitions, and they may have difficulty breaking down a vocabulary word into different parts. This makes understanding and interacting with the vocabulary prior to reading an essential part of the prereading process.

The Frayer model (Buehl, 2017; Frayer, Frederick, & Klausmeier, 1969) is a simple, adaptable strategy designed to help students demonstrate their understanding of complex vocabulary in a simple, adaptable graphic organizer. This helps students adjust or better understand the concepts in isolation and then work them into the context of the reading and any subsequent writings.

How to Use

The Frayer model graphic organizer identifies a vocabulary word and then asks students to address four parts to aid them in comprehending that word: (1) what it is (a definition), (2) what it is not, (3) characteristics, and (4) examples and images. As we show in figure 3.3, the Frayer model is adaptable to fit a variety of reading objectives. (See page 180 for the reproducible "Frayer Model Template.") The graphic organizer format allows readers to deconstruct a vocabulary concept into smaller parts to gain a broader understanding of the word or social studies concept. By breaking the word down into different parts, students draw connections between different concepts and identify gaps in their understanding. It is often successful in a jigsaw manner, where students take on different words and then share their work in small groups (Aronson & Patnoe, 1997).

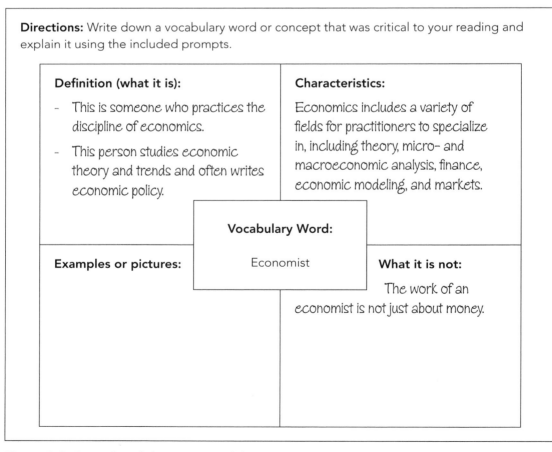

Figure 3.3: Examples of the Frayer model.

One way teachers can use the Frayer model for social studies vocabulary is to have students skim through the learning targets (process standards) for a unit to find words they do not know. This also personalizes the learning for each individual by allowing students to individualize their vocabulary acquisition; proficient students won't spend time reviewing words they already know, and students with less background knowledge can bridge gaps to new learning. After the prereading, our team of literacy experts and social studies teachers developed a few short activities to have students share or check the knowledge they obtained and the words they learned. Figure 3.4 highlights the steps for this process.

Day 1: Students Determine Important Vocabulary Words (Students should complete this independently after the instructor explains or models how to complete this activity.)

1. Students read targets from the unit of study and think through the targets to focus their understanding of purpose.

2. The teacher models using the Frayer model graphic organizer by skimming for and reading through the section on a specific term. As part of this modeling, students should:

   - Read the definition. This information can come from the sources provided for them, the internet, class discussion, and so on.

   - Reflect on their understanding of the term if they understand how it is defined in the text. If one or more students do not understand, have them consider how they might identify other sources to clarify the meaning.

   - Based on the definition, fill in on the Frayer model a description of what it is not.

   - Skim the reading again and select characteristics that apply to the selected term.

   - Draw an example using the visuals in the text.

3. The teacher assigns homework based on what he or she feels is the next step in understanding the concepts.

Day 2: Students Engage in Optional Short Activities to Check Understanding and Address Targets (This is a postreading activity.)

1. The teacher has students compare graphic organizers for accuracy. (Students can use their specific graphic organizers from completing the activity in class the previous day or from completing it as homework.)

2. The teacher has students compare and contrast various combinations of the terms in their graphic organizer. (This is a collaborative activity with classmates, but the instructor should model it.)

Figure 3.4: Process for applying the Frayer model.

*Visit **go.SolutionTree.com/literacy** for a free reproducible version of this figure.*

Adaptations

You can adapt the Frayer model for students who are struggling with the literacy aspect of the strategy to help them better understand the content. This strategy helps by creating a web of knowledge around a central idea or specific vocabulary words that could be challenging for these students to understand. Using the Frayer model helps students remember, access, and clearly identify specific concepts; but for these groups, teachers should adapt the prompts in each box for accessibility. The answers and learning for *all* students, however, should be the same. Teachers can also provide hints or sentence starters to help students in special education. When you consistently use the strategy in this way, students struggling with social studies terms become more familiar with the process and how to implement it to aid in their understanding.

In addition to these ideas, for students learning English, filling out the Frayer model boxes in both their primary language and in English can further assist with bridging gaps and providing a clearer understanding of the concepts. Another adaptation to consider when introducing the strategy is to provide a model for the students and create a matching game between the vocabulary words you want to introduce and their corresponding definitions. Providing students with such a model clearly outlines expectations for their work and allows social studies students to interact more with the words.

In addition, some students may exceed the knowledge base and need a more challenging task. You can adapt the Frayer model strategy by changing the tasks within each of the boxes to demand higher levels of thinking. For example, teachers might ask students to provide an example and explain how the example fits the definition using evidence. This is part of what makes the Frayer model so adaptable for all students.

Visuals

Accessing images before reading can help students comprehend and gain important context about a text that will help them during reading by creating mental images that connect with a text and make it come alive. Often, strong readers already create mental images as they read and have a movie playing in their mind of what they read in the text, whereas struggling or reluctant readers sometimes have a difficult time creating images derived from words on the page. This strategy helps readers form such mental images, and research shows that readers benefit from direct instruction around imagery and reading (de Koning & van der Schoot, 2013). Specifically, when teachers provide images for students as a prereading activity, they support students in building a visual representation of text. In other words, when students can see images prior to reading a text where they may lack prior knowledge, the images will better help them visualize (play the movie in their heads) while reading.

How to Use

As the teacher, find and display visual information that makes the text come alive. You can project or distribute (as handouts or electronic files or links) these images for a discussion before reading or engaging with a text. They will then assist students in making mental images while reading. Questioning and guided visualization based on the images further provide students with mental images to assist in achieving clarity of what the words in the text say. Prior to reading, the images provide students with visuals of events they may read about. For example, if your students will be reading a text about World War II, you might use a resource, such as Good Free Photos (www.goodfreephotos.com/historical-battles/world-war-ii), to collect a series of images related to the reading and present them to students.

Adaptations

This strategy is very powerful when working with students who struggle with reading comprehension. Providing concrete images to students breaks down language barriers and helps them make better connections to the content of the text. Providing many visual examples, such as pictures, graphs, infographics, signs, and so on also supports meaning making with the written text (Allison & Rehm, 2007).

Inquiry Scenarios

Inquiry scenarios provide students with a series of short readings that expose them to important questions that they will explore throughout the subsequent reading of the primary assigned text. Using this approach as a prereading strategy helps build students' background knowledge about a concept they are going to engage with by having them assess, build, and use knowledge that is consistent with the concept (Wilhelm, 2007). With this background knowledge established before the primary reading, students can tap into that knowledge during reading and use their critical-thinking skills to make decisions, solve problems, and explore issues they may be reading about. This process also allows students to form initial opinions about a topic from the prereading activities and then use the during-reading content to confirm or change their opinions, as appropriate.

Social studies teachers want students to see the historical context surrounding a reading, as well as find it purposeful and potentially connected to their own lives today. By asking students to consider these big questions before reading, students are better equipped with a strong foundation around the social studies topics you select.

How to Use

A basic scenario for this strategy can be using a text to create a situation that students need to consider and apply course concepts to. For example, a class preparing to read about Pearl Harbor might explore the inquiry question, When is a nation justified in going to war? To dive into this question, students might read and discuss several texts that provide perspective about the nature of conflict. One text might explore a baseball game where two teams are on the brink of fighting due to one or more events in the game. Another scenario might cover a young person debating whether or not he wants to join the Army. Similarly, students can read a scenario about a teen arguing with his or her parents about attending a war protest. The scenarios don't provide concrete answers. Instead, they provide students with a port of entry to discuss some of the pro and con arguments used to support the war, setting them up to be more successful when they engage in rigorous reading about Pearl Harbor and the United States' subsequent actions.

Giving students inquiry scenarios encourages them to ponder specific concepts before reading and interpreting a text. This strategy can be as formal or informal as the teacher prefers. For example, a teacher can display the scenario for the full class and have students write a response to the inquiry question within the context of the scenario. Alternatively, he or she might put students in pairs or small groups, ask them to think about the scenario, and then brainstorm how it will play out. The teacher might even have students act out scenarios, with students adding conclusions they feel are best for the scenario. Students might also draw on their own experiences in connection with the scenario to consider their own positions.

It is always beneficial to revisit the scenario after students have completed the prereading and during-reading activities. This will confirm their thought processes and validate their predictions of the reading or encourage them to reflect at a high level to revise their own thinking.

Adaptations

This strategy is very powerful when working with students who qualify for special education. Incorporating elements of the Visuals strategy (page 66) by providing several visual examples—such as pictures, graphs, infographics, signs, and so on—supports meaning making with the written text and helps them access more information than if they had no background at all (Allison & Rehm, 2007). In addition, you might have some potential response to the scenarios so that students do not spend too much time off task if they have trouble coming up with a response.

For students learning English, providing practice situations that might be more relatable for the students to explore will help break down language barriers and help them make better connections to the content of the text. Also, going over the scenarios orally before students start working on them helps generate a clearer focus on the task.

When using inquiry scenarios as a prereading strategy, let all students know it is OK to *not* have the correct answer or know the correct outcome. Articulating this to the students is key, as it helps create a more comfortable, low-risk learning environment for all. Even students who can already demonstrate proficiency with this strategy benefit from experimenting with low-risk activities, such as this one.

Beliefs and Opinions Survey

A beliefs and opinions survey serves as a type of anticipation guide that gets students thinking about concepts in the text. Beliefs and opinions surveys ask students to agree or disagree with statements about ideas in the text prior to reading it. Because there are no right or wrong answers in the prediction phase of this strategy, reluctant readers are often more comfortable participating in this activity. Students with little to no background knowledge about the unit of study can explore their ideas and beliefs about the subject prior to reading. By exploring their own ideas before reading, students help to activate prior knowledge, preview the ideas of the text, and set a purpose for reading. Once students have an opportunity to consider their own beliefs and opinions, they can begin to explore what the text has to say about the same questions.

How to Use

Prior to the reading, the teacher creates multiple meaningful statements about thematic topics that students will engage with in the text and formats the statements in a survey that asks students to mark if they agree or disagree. It may be helpful to provide students with a scale to demonstrate how strongly they agree or disagree with a statement. This allows students to think about some of the nuances of each statement instead of offering a blanket opinion. Another option, one that gets students up and moving, is to have them stand in a line with those who strongly disagree with a statement on one end and strongly agree with the statement on the other. Encourage students who are comfortable doing so to share their rationale for their position.

Next, students read the text. After reading, they go back and respond to the survey again, considering whether they still agree or disagree with the statements or if they've changed their minds. They can then write down why they have changed their beliefs and opinions and provide evidence or quotes from the text to support their answers. Figure 3.5 provides an example for a social studies classroom preparing to read about amendments in the U.S. Constitution. (See page 181 for the reproducible "Beliefs and Opinions Survey.")

Read the following statements and circle your reactions. Be prepared to defend your opinions with evidence from your own experiences.

I believe all laws should be strictly enforced at all times.

Strongly Disagree	Disagree	Neutral	Agree	Strongly Agree

I believe that every U.S. citizen should have the right to own a gun for protection.

Strongly Disagree	Disagree	Neutral	Agree	Strongly Agree

I believe that everyone has a right to free speech no matter what.

Strongly Disagree	Disagree	Neutral	Agree	Strongly Agree

I believe that the death penalty is sometimes necessary depending on the crime.

Strongly Disagree	Disagree	Neutral	Agree	Strongly Agree

I believe that police should have the right to search property, including cell phones, without a warrant.

Strongly Disagree	Disagree	Neutral	Agree	Strongly Agree

Figure 3.5: Beliefs and opinions survey example.

Adaptations

When adapting an opinion survey for students who qualify for special education, it may be more effective to work with one statement at a time. The teacher can read the statement aloud and ask students to move to different places in the classroom to indicate their positions. This helps focus students and not overwhelm them with multiple statements all at once. Teachers can move on to subsequent belief statements if more time is available after discussion of the first statement.

For students learning English, it may be worthwhile exploring how cultural norms influence beliefs. It is valuable for students adapting to a new culture to see how different members within a culture respond to different ideas and beliefs. This activity also allows the teacher and the students' own peers to explore how cultural values may impact the reading of a text that may represent unfamiliar culture norms. Given that students may feel vulnerable exposing their cultural background, the classroom community must have well-established norms for these discussions. Consider using a resource such as Teaching Tolerance (teachingtolerance.org) to access suggestions for ensuring the classroom environment is a safe place for students to share.

For proficient readers, particularly if you have a large number of them in your class, a good follow-up is to have students generate their own belief statements. Teachers can collect them and aggregate them into one document where the class can respond. It's important to keep the statements anonymous. This move to student-generated material is a logical next step for students who have mastered the strategy.

Turning Titles and Headings Into Questions

Setting a purpose for reading is essential when preparing students for a successful reading experience. Authors and educators William E. Blanton, Karen D. Wood, and Gary B. Moorman (1991) and Cris Tovani (2005) assert that establishing a clear purpose for reading guides the reading process, activates background knowledge, provides a common focus, and helps students to determine where to focus their attention. Using this strategy, students consider a text's title as well as any chapter titles or section headings (these tend to denote a text's main ideas) and reframe one or more of them as questions. This helps students more thoughtfully analyze the text's purpose before reading by honing in on key details that will support their reading and ability to identify main ideas. Ultimately, students will explore the reading with the expectation of finding answers to their questions.

How to Use

Students should go through the text and identify all of the headings in the reading. For example, this might include headings within a single chapter or sections within an article. For each heading, the student should rewrite the title as a question. For example, consider the following headings and subheadings in a world history unit on the Renaissance.

- The Renaissance in Italy
- Rebirth of the Classics
- The Renaissance Is Born
- Marco Polo
- Florence: A Renaissance City

The student could take notes on the reading by organizing notes around questions created using the words in the headings.

- What and when was the Renaissance in Italy?
- Why and how was there a rebirth of the classics?
- Why was the Renaissance born?
- Who was Marco Polo?
- Why is Florence considered a Renaissance city?

For each of these student-generated questions, students can take notes on key details that answer the questions. It is a simple way to reframe the reading by setting a clear purpose for reading.

Adaptations

This strategy is powerful for students who struggle because it sets their purpose for reading if they don't know a lot about the topic or where to start their reading. Students who struggle with this literacy strategy might need extra modeling of questions and help with creating questions. Additionally, you might place students in groups with different ability levels to help students who need skill development to practice the strategy with students whose skills are currently more adept.

In addition to purposeful grouping, English language learners might benefit from teachers providing a word bank or question stems as they practice the skill. Depending on their level of English proficiency, their primary language may structure questions differently, so students may need several examples of ways to structure questions as they work on mastering this skill.

Like we mentioned previously in this section, it can be useful for students who excel at this skill to work in small groups helping other students to also master it. Additionally, these students can try to write multiple possible sentences for each title to expand their focus later while reading as they look for answers to their additional questions.

Possible Sentences

Students are more likely to thrive with a rigorous text when the teacher has made every effort to activate the previously discussed four Ps of prereading. This strategy allows students to make predictions about what it is they are about to read by using a teacher-provided list of ideas related to the reading and then writing a sentence related to each of them. The idea is that students anticipate sentences that could possibly appear in the text. In doing this predictive work, students activate their own prior knowledge about the text, and they set a purpose for the reading that is to follow. Although we categorize this as a prereading strategy, it can also serve as a during-reading and postreading strategy.

How to Use

Provide students with a list of important terms, concepts, and names from a reading. For each item on the list, have students write a sentence that features it. Encourage students to preview or skim the reading and rely on prior knowledge when writing their predictive sentences.

During reading, students read to analyze the accuracy of their predictive sentences. Teachers may prompt students to underline details in their sentences that need revision based on new or clarifying information they encounter in the text.

After reading, students should revise any inaccurate sentences with correct or missing information. This allows students to reflect on their new understanding of the reading by writing revised sentences in their own words.

Adaptations

As with all strategies, possible sentences offers valuable scaffolds for students who are struggling with a reading, especially when the teacher takes time to effectively model this strategy. Such modeling shows students how to interact with the vocabulary, ideas, concepts, thoughts, and so on, and supporting this with examples provides students with an outline of how to put the words together.

For students learning English, in addition to modeling, it may be necessary to provide a glossary or change certain words to make them easier to understand. Since many other languages structure sentences differently, these students will benefit from models and practice. Treat any simpler word variations you offer as scaffolds intended to bring students up to full grade-level vocabulary. Given the varying syntax of other languages, reassure students who struggle that it is okay for them not to be 100 percent accurate when writing predictive sentences.

Encourage students who show proficiency with this strategy to make an increasing number of connections between what they predicted and what they discovered from the reading. Sustain their engagement through the reading process by emphasizing the fun to be had in predicting content as accurately as possible.

Picture Books

Many middle or high school social studies teachers might scoff at the idea of bringing a children's book into their classrooms. Considering that the CCSS encourage teachers to increase the rigor of texts, it may seem counterintuitive to have students interact with a picture book geared toward beginning-level readers. However, a simple and accessible text may allow for instruction that isolates a specific reading skill or thematic topic. To build thematic connections, teachers can match picture books to the upcoming grade-level reading to increase vocabulary, build prior knowledge, and create engagement. This approach is distinct from the Visuals strategy (page 68) because picture books are usually connected to a narrative or theme in some way. This strategy helps students better understand the purpose of the learning. For example, a set of picture books based on race relations may set the stage for a reading about Jim Crow laws. Or, if a teacher wants students to pay more attention to visual information in their reading, a picture book without words or with minimal written text may help students focus on images to make meaning and inferences from subsequent visuals they might encounter while reading a more rigorous text.

In these ways, picture books provide opportunities for students to explore thematic concepts and concrete reading strategies that students will encounter in subsequent curricular texts. By isolating the theme or strategy in a simpler text, students can have practice opportunities without the challenge of comprehending a rigorous text. In turn, they should be able to apply those strategies and deeper understandings of themes to more challenging texts they will interact with in the future. In the past, we have worked with school and local librarians as well as used online searches to find suitable books to pair with thematic topics or skills.

How to Use

Teachers ask students to perform a specific task. For example, if all students have copies of the book, ask them to place a sticky note with a prediction near an image or text that helps give them an idea about what will happen next. The idea is to encourage students to use their understanding of the topic to guide their predictions based on what they see and read. No matter a student's reading skill, the picture book should encourage him or her to grapple with points of inquiry and skills that are similar to what the student will need to do in more complex reading tasks. Guidelines for how students should engage with the text, and for tasks such as the sticky note exercise, may include the following.

- ▸ Make predictions.
- ▸ Find examples of figurative language or symbolism.
- ▸ Identify a key detail related to character or plot.
- ▸ Identify a theme.

▸ Make personal connections or build on prior knowledge.

▸ Develop vocabulary by taking note of unfamiliar words.

If it's not possible to obtain multiple copies of the children's book you want to use, implement this strategy as a whole-class activity or place a limited number of copies at a station as the class works on other prereading strategies. Students can record their thinking in several ways: on the board, on a poster, or with a simple T-chart graphic organizer that records evidence from the text in one column and analysis in the other. Another option is to allow students to have discussions about their findings, perhaps recording key takeaways in their notes while forgoing the sticky notes.

As an example of how this might look during a U.S. history class studying recent events, students would connect recent examples of racial strife in the United States with similar events during the civil rights movement of the 1960s. As students grapple with issues of racial inequality throughout U.S. history and the circumstances that may have led to protests and riots, the teacher may have the class read the children's book *Smoky Night* (Bunting & Diaz, 1994). In this picture book, Daniel, a young African American boy, observes people in his neighborhood acting out, sometimes violently, after the acquittal of four white police officers accused of beating Rodney King, a black man. A teacher might ask students to place sticky notes anytime they encounter examples of anger at racial injustice and its effects on the local community, a topic they would encounter during the unit as they look at various ways people worked to gain civil rights in the United States. Students would be able to share their findings, and the class can begin to address many of the thematic issues that it will encounter in more rigorous texts later in the unit.

When the class begins to engage with more rigorous readings related to the civil rights movement, the teacher can refer to students' experiences with the simpler text. This gives all students, regardless of ability, the opportunity to practice a skill while developing a deeper thematic understanding.

Adaptations

Students with reading abilities that are significantly below grade level may benefit from having audio versions of the children's book to assist with decoding or fluency issues. If the text is at a level where students can read independently, students can read the book aloud to a partner for practice with decoding or fluency skills as they work their way up toward grade-level reading. We believe it is best to isolate one skill or theme for students to focus on so they have a clear purpose for reading.

For students learning English, picture books may provide opportunities to address cultural knowledge, provide historical background, or introduce thematic vocabulary they will explore in the more challenging texts that follow. The illustrations can serve as a powerful visualization tool for students reading about unfamiliar content or settings.

In a classroom with proficient readers, teachers can vary the task to meet the abilities of the learners. For example, as opposed to only identifying the theme, have students reflect on other text-to-text, text-to-world, or text-to-self connections where they've encountered similar or contradictory themes. These reflections can be written, oral, or put into questions they can share with other students in the class.

Considerations When Students Struggle

The reality is that some students will struggle to use and master prereading strategies and the four Ps of prereading. Such struggles may occur for many different reasons, from not being able to recall prior concepts to not understanding new concepts to not being able to make connections between concepts. Here are some additional questions to consider when you encounter students who are having a hard time applying prereading strategies.

▸ **How might you ensure students are regularly stopping and thinking about a text before reading?** To answer this question, it may be helpful for the social studies teacher team to do a prereading inventory. Before beginning a new unit or embarking on a text, consider the different ways that teachers on the team have provided entry points for students to prepare them for the rigorous reading that awaits them. Teams may find there are already a variety of instructional tools in use. In other instances, there may be evidence that the team can do more to set students up for success with texts using some of the strategies in this chapter.

▸ **How might you provide students with additional prior or background knowledge they may need before reading? What resources are available in your building to close knowledge gaps?** Teacher teams can work together to anticipate difficulties with specific texts. If the team has taught a text before, teachers likely already know several things about where students struggle. The team can collaborate around ways to bridge gaps to help students with understanding before they begin the reading.

▸ **How might you encourage students to make predictions before reading?** Teacher teams can share strategies that they use before reading something for the first time. Generate a list as a team, and allow team members the freedom to experiment with different prediction strategies. Discuss as a team the successes and challenges that these experimentations result in.

▸ **How might you help students understand why they are reading?** Students need to see a clear purpose for their reading. When your team is planning a unit, is the team finding ways to ask big questions that make the reading relevant to students' lives? Are there points of inquiry throughout units that allow students to explore multiple layers of texts?

Note that your answers to these questions can be highly variable depending on the nature of your school, what you are working on with your students, and so on.

Review the strategies in this chapter. How might you help students who struggle to make progress?

thinking
BREAK

Considerations When Students Are Proficient

For students who have demonstrated proficiency, teams should collaborate around prereading strategies that help students get to a deeper meaning of a text. Here are some additional ideas for proficient readers in the prereading process.

▸ Encourage proficient readers to share their prior knowledge and make connections with other proficient readers and their prior information of the text. By sharing this information in a partner group, proficient readers will be able to extend their thinking and ideas and express these with like minds.

▸ Assign proficient readers to work with nonproficient readers during the prereading process so the former can share their understanding of the concepts with the latter. In helping relate their understanding of a text, both proficient and nonproficient readers benefit from this collaborative work.

> ▸ Provide extended choice opportunities to proficient readers (in addition to the choices you provide to all other students) that push them to higher thinking levels. This reading material can include higher grade-level or college-level reading material. In implementing this practice, do not provide selection opportunities exclusively to highly proficient readers. The message should never be that only proficient readers are entitled to choice or additional learning opportunities.

You will note that for many of the strategies in this chapter, we acknowledge equal effectiveness for proficient and struggling readers. Every student, no matter their existing proficiency level, brings something unique to every text, and teachers should celebrate those differences in their classrooms. Students have such varied experiences and backgrounds that sharing their unique prior knowledge can create an exciting environment for learning, while also showing students that their backgrounds and experiences, no matter what they are, have value. When there are opportunities to push proficient readers, teachers should do so; consider some of the different levels of questioning addressed in chapter 5 (page 107). Bear in mind, though, that all students should have exposure to prereading instruction.

Wrapping Up

Using the prereading strategies in this chapter will help students activate their brain glue (schema) and be prepared to take in essential content information. You can implement all of these strategies in different classrooms and adapt them for all students to use. Recall that all students should know how to preview a text, make predictions about what they are reading, activate prior knowledge to make connections with texts, and see a clear purpose for all reading tasks. When these practices become part of the classroom routine for all reading tasks, they become ingrained in students' reading habits. When teachers argue that there isn't time for prereading, we argue they are setting up their students for frustration and potential failure. We owe our students the time to prepare them to be successful with all reading tasks. Always refer back to your four Ps, ensuring students understand how to preview, predict, use prior knowledge, and read with purpose.

Collaborative Considerations *for* Teams

- Which strategies are best suited for your students? Consider the following.

    - Your students' needs

    - The experiences and knowledge they bring to the reading

    - The experiences and knowledge they need in order to understand the reading

- Which strategies are best suited for the text and accompanying tasks?

- How can you differentiate these strategies for the different learners in your class?

- How can your team work together to create and use strategies that best teach all students?

- Which strategies are best suited for the targeted C3 Framework for Social Studies Standards?

- Does the team have access to student data that will inform it as to which strategies are appropriate for specific groups of students?

CHAPTER 4

During-Reading Strategies

Even after deploying sound prereading strategies to lay the groundwork for a reading, students need additional strategies to fully engage them in the primary grade- or course-level text. In our work with social studies teachers, the more our literacy team collaborated with them, the more we heard about students not completing assigned reading competently, if at all. In fact, a faculty survey during the 2013–2014 school year about teacher perceptions on student literacy revealed our social studies teachers believed that 38.5 percent of students fail to complete assigned reading. We discuss the reasons for this outcome later in the chapter, but ultimately, we realized we had to find a way to infuse the classroom with during-reading literacy strategies and activities while staying true to the course content to help students succeed and grow as both readers and social studies students.

This chapter explores how social studies teachers can work together as a team to select during-reading strategies that support students as they read literary and nonfiction texts. After looking at the need to build engagement and foster active reading, we review the C3 Framework for Social Studies and how it aligns with the CCSS ELA to highlight embedded during-reading literacy skills in the standards. With this knowledge in place, we offer a series of concrete during-reading strategies we have developed through team collaborations with social studies teachers. The chapter concludes with considerations for addressing students who continue to struggle and those who show higher-level proficiency.

The Need for Engaged, Active Readers

Try to remember the last time you assigned a reading task to your students. Did they respond with enthusiastic high fives and fist pumps, or did they shrug with a general sense of apathy? For many educators, the latter experience is the more likely student reaction when faced with a new reading task. The students most interested in the course content will surely get through the reading, other students may skim through the text to ensure they have a sense of the main idea, and still others rest their heads on their hands, quietly reading line by line but not truly engaging with the text. Most critically, as our survey at the start of this chapter revealed, most students will fall short of completing the task.

Imagine your quiet, compliant students reading. Now compare this image with one of yourself at a time when you were fully engrossed in a text. Perhaps you recently found a book or article that you found particularly fascinating. When you become fully engaged while reading, you leave your current surroundings to enter the world of the text. You tune out everything around you and begin to place yourself in the moment. Psychologist Mihaly Csikszentmihalyi's (2009) research reveals that this isn't limited to literary texts; reading an article about current events or important ideas can just as easily pull readers into a state of *flow*, a level of engagement where one becomes completely enveloped in a task or experience. Visualize your own body language when you are completely engaged in a text, and juxtapose that with the image of your students reading with heads weighted on hands. How do you make that image of your students look more like your own, including for those students who wouldn't include a social studies class on a list of things they're passionate about?

Linda Gambrell (2011), a former president of the International Reading Association, argues that students are motivated to read when they have opportunities to be successful with difficult texts. Gambrell (2011) also asserts when students have a clear sense of not only the *what* but the *why* behind a reading task, they are more likely to be engaged and find meaning in their work. By taking the time to activate prior knowledge and set a purpose for reading, you are already on the right track toward creating a more engaging, active reading experience for your students. With this foundation set, when teachers ask students to make meaning with rigorous literary and nonfiction texts while also ensuring they have the tools they need to be successful, those students take an increasingly active role in their learning.

thinking
BREAK

Consider your most recent reading assignment. How did you set it up for your students? Did you give them a specific task or role? Did you simply tell them what pages to read and when the reading was due? What did you expect students to do while reading that will keep them active and engaged with the content? How might your use of prereading strategies from chapter 3 (page 45) help establish a stronger sense of purpose for during-reading tasks?

Collaboration Around During-Reading Activities

We wrote in the introduction for this chapter that a running theme in our collaborations with social studies teachers is students not completing assigned reading. When our social studies collaborative teams, working within our school's PLC, gathered to discuss and explore the potential reasons for this low reading-completion rate, they identified four factors they believe contribute to this trend.

1. Students struggle to engage with complex texts.

2. Students fail to connect to texts that they see as irrelevant to their lives.

3. Students become accustomed to relying on teachers to summarize reading for them.

4. Electronic resources that provide text summaries and overviews of social studies concepts are easily accessible and enable students to seem like they are reading when they aren't.

During team discussions, the frustration was palpable because we knew that students would not learn and grow if they did not read. So, our collaborative teams had to adapt and change to meet students where they were and use strategies that would engage them during the act of reading. The team acknowledged that if students were going to take ownership of their reading, we needed to teach them how to engage with texts the same way that a social studies expert does. It sounds simple, but this was a turning point for the team, realizing that the answer to the first critical question of a PLC (What do we want students to learn?; DuFour et al., 2016) was that our team wanted students to deeply engage with texts. We wanted students to embrace the struggle of reading, learn how to make meaning during the act of reading, and fully engage in the reading process.

For many teachers, this can be a daunting task, especially if they don't have concrete strategies to utilize with students. If your team doesn't know where to begin, then begin with one strategy from this chapter and build from there. We believe the strategies present in this chapter give teams an inroad to during-reading instruction to help students deeply engage with the varied texts of a social studies classroom.

C3 Framework for Social Studies Connections

The CCSS ELA require students to read increasingly complex materials at a high level, a requirement that applies to nonfiction social studies texts (NGA & CCSSO, 2010). To be college ready, students need to have strategies at hand to analyze and break down texts—especially the nonfiction texts that will be a staple of their academic and adult lives. As part of a social studies teacher team, many of the ways that your mind operates when engaging with a text are second nature. However, for many students, this does not come naturally. Your role (and your team's role) is to provide instruction for your students that makes these thinking habits explicit and helps your students make meaning of texts during reading, as required in the CCSS for ELA.

As we discussed in the previous chapter, there is a direct connection between the four dimensions in the C3 Framework for Social Studies and the CCSS ELA skills. So, as in chapter 3 (page 45), table 4.1 provides potential connections between the strategies presented in this chapter and during-reading skills in the CCSS Anchor Standards for Reading. Table 4.2 does the same using the four dimensions in the C3 Framework for Social Studies. (See the C3 Framework for Social Studies Connections section in chapter 3 [page 45] for a detailed list of the Anchor Standards for Reading.) Note that, with these tables, we are purposely checking several standards that any given strategy may fulfill. While not every strategy connects directly or initially to the standards, teachers can use their professional knowledge to adapt and tweak a strategy to address the targeted standard or standards.

Table 4.1: CCSS ELA During-Reading Strategy Connections

During-Reading Strategy	R.CCR.1	R.CCR.2	R.CCR.3	R.CCR.4	R.CCR.5	R.CCR.6	R.CCR.7	R.CCR.8	R.CCR.9	R.CCR.10
Read Aloud, Think Aloud	✓	✓	✓	✓	✓	✓		✓		✓
Find Evidence to Support Claims	✓	✓	✓	✓				✓		✓
Text-Dependent Questioning	✓	✓	✓	✓	✓	✓	✓	✓	✓	✓
Primary Source Graphic Organizers	✓	✓	✓		✓	✓		✓	✓	✓
Text Chunking	✓	✓	✓	✓	✓	✓				✓
Lateral Reading	✓	✓	✓	✓	✓	✓	✓	✓	✓	✓
Building Reading Stamina	✓	✓	✓	✓	✓	✓				✓

Source for standards: NGA & CCSSO, 2010.

Table 4.2: C3 Framework for Social Studies During-Reading Strategy Connections

During-Reading Strategy	Dimension 1: Developing questions and planning inquiries (R1)	Dimension 2: Applying disciplinary concepts and tools (R1–10)	Dimension 3: Evaluating sources and using evidence (R1–10)	Dimension 4: Communicating conclusions and taking informed action (R1)
Read Aloud, Think Aloud		✓	✓	
Find Evidence to Support Claims		✓	✓	
Text-Dependent Questioning	✓	✓	✓	
Primary Source Graphic Organizers	✓	✓	✓	
Text Chunking		✓		
Lateral Reading	✓	✓	✓	
Building Reading Stamina		✓		

Source for standards: National Council for the Social Studies, 2017.

Strategies for Supporting Students During Reading

If you've utilized prereading instructional strategies, your students are already better equipped to engage with texts in meaningful ways. To create a reading experience that is active and engaging for social studies students, consider the different ways during-reading instruction can build on the prereading skills in chapter 3 (page 45). For example, once students have a clear intent for reading, they are more likely to be purposeful in their during-reading annotations. Likewise, if students have activated prior knowledge relevant to a text, they are more likely to make text-to-self connections or visualize the text. Essentially, prereading strategies help students to understand the *why*, making it a natural progression for during-reading strategies to help establish the *how*. For example, if students used the turning titles and headings into questions strategy (page 71) to clarify the purpose of a reading about the American Revolution and they come across a section titled "The Basic Causes of the Revolution," they already know to look for those causes from their effort to create a question about it (perhaps, *What were the basic causes of the American Revolution?*). They must then create a list of factors from the full reading that answers the question. In essence, the question becomes the purpose, and the *how* is identifying a list of factors through annotation or a graphic organizer.

All the strategies in the following sections are designed to turn students from passive participants into active thinkers. Instead of merely getting through the reading, students navigate their way to new understanding. Along the way, these strategies encourage them to question and challenge a text, embrace uncertainties, and engage in a dialogue with the text. As part of teaching and modeling these strategies for students, you can—and should—share your passion for history and social studies texts with your students. Connect that passion for the world to the process of learning about it. For example, if students see how you read with passion and enthusiasm about an event to understand its causes and effects because you know it will help others not to repeat the mistakes of the past, you will nurture and grow that same passion within your students. When you explicitly teach students to think like a passionate historian by modeling the reading strategies in this chapter, they are far more likely to find themselves engaged in new learning.

Read Aloud, Think Aloud

The read aloud, think aloud (RATA) strategy is geared toward promoting metacognitive reading habits. Most teachers can probably identify students who struggle with reading rigorous texts, but understanding *why* a student is struggling requires a closer examination of his or her reading habits. This strategy encourages readers to verbalize their thinking while reading, which can, in turn, help teachers differentiate instruction that better supports students' literacy and disciplinary growth.

Consider how much goes through a reader's mind as he or she constructs meaning or struggles through a text. Readers constantly decode words, ask questions, clarify points of confusion, make predictions, and make connections to other social studies information they know—just to name a few! RATA helps students monitor that thinking and establish mindful habits while reading by making that reading and thinking transparent. When students articulate their thoughts from a reading, they not only clarify their own sense of the text, they also help illuminate where comprehension is breaking down.

How to Use

Put students in small groups of two to four, and assign each group member a small portion of the text. The smaller the group, the better. Students take turns reading aloud portions of the text, frequently stopping to clearly articulate their thinking while their peers annotate their own text. Annotations should reflect their thinking, so they may vary from student to student. Students may write questions, establish connections, craft summaries, and so on.

At first, students may need help identifying what to share as they read. Through modeling, encourage them to make connections with prior knowledge, identify areas where they are confused, make connections with bigger ideas in the unit of study, question the text or author, or share any other ideas that run through their minds as they read their assigned portion of the text. As they read, group members listen and annotate their own text. Annotation instructions can vary, so the teacher should be clear about what students are marking and why based on the nature of the text. For example, teachers can encourage students to ask questions, note confusing aspects of the text, make connections, and so on. It's crucial for every student to have a purpose for reading, and annotations help them to be accountable to that purpose. When the reader is finished with the assigned portion, have group members contribute their thoughts (using their annotations as a guide), perhaps adding new insights or clarifying points of confusion the reader had.

To help sustain engagement, it is best to assign shorter excerpts—that is, passages that will take no more than five to ten minutes to read—for each student. A common question we get from students about the RATA strategy is, "Will this slow me down?" Yes! That's the point. You want to provide students an opportunity to become more aware of their own thinking when navigating a challenging text, and you want them to have an opportunity to hear how others think when reading one as well. For this reason, we encourage social studies teachers to briefly model the process to demonstrate how a literary expert thinks when reading. Teachers should "perform" this for the class by reading the passage aloud and narrating their thinking, adding annotations as they read. Several postreading strategies make an effective follow-up to the RATA strategy (see chapter 5, page 107), but at a minimum, have each group share its biggest takeaways with the whole class. As students think aloud in their small groups, circulate throughout the room to observe closely. Consider taking notes on common patterns or key insights from students and provide formative feedback to improve their approach to reading.

We understand that for many social studies teachers, providing time in class to read may seem unusual or even impractical, but consider the benefits of observing students as they articulate their thinking and insights using this strategy. This is a powerful formative experience that allows your team to make meaningful adjustments to instruction, as well as provide students with meaningful feedback that can help them achieve higher-level thinking.

Adaptations

For students who have reading or speech difficulties, including students learning English, reading aloud may cause anxiety or nervousness. For this reason, avoid conducting whole-class RATAs, as these can cause even proficient readers to focus more on recognizing words than on constructing meaning (Worthy & Broaddus, 2001). This is why we encourage the use of small groups, which also allows teachers to organize students into similar ability groups or into groups of deliberately matched peers at different levels. The latter of these provides an opportunity for proficient readers to share expertise and take on a leadership role. Both arrangements can build confidence and raise reading fluency, and they create a more comfortable environment for students when they are working with peers of similar ability or those they can trust to support them. We contend that when students who are struggling have the opportunity to hear how their peers construct meaning and navigate rigorous texts, they see that it is also possible for them to be successful and that deep reading is not "just a teacher thing."

Also, ensure students learning English or others who may struggle to read a text have an opportunity to read their section to themselves prior to reading aloud with their group. This first independent read allows students to develop some familiarity with the text, so they aren't worried about how to say the words. You can also provide students the opportunity to ask their group members how to pronounce some of the more challenging words or discuss difficult disciplinary concepts. Adapting this part of the strategy, along with the actual task of reading aloud, further assists students to succeed.

Find Evidence to Support Claims

When students have a clear purpose for reading, they know what they are looking for in a text. Building on that idea, consider the benefits of providing students with several claim statements prior to an assigned reading. Students review the claim statements before reading, perhaps in tandem with an anticipation guide (see page 56), and then they search for evidence that either supports or refutes the given claims while reading. Students weigh the details in the text against important ideas and points of inquiry.

How to Use

Consider important themes the class will explore when interacting with a text, and generate a list of thematic claims that the text addresses. These should be important statements related to the big ideas of the unit or text and not necessarily

something the text proves or disproves. Instead, include contradictory or opposi-
tional claim statements that allow students to consider different interpretations.
With a variety of claims to evaluate, students consider and discuss their own position on a theme or claim statement, and then they search for evidence related to
the claim as they read. The subtle difference between this during-reading activity
and the anticipation guide (see page 56) as a prereading activity is that the claims
for this exercise should be arguable, whereas statements in an anticipation guide
tend to be true or false.

Figure 4.1 shows an example in which a student has recorded textual evidence
while reading and is considering whether the evidence supports or contradicts the
claim. (See page 182 for the reproducible "Find Evidence to Support Claims.")

Finding Evidence: Witchcraft in Salem		
Thematic Claim Statement	Is Claim Accurate?	Evidence to Support or Refute the Statement (include page numbers)
Tituba and her followers were charged with witchcraft due to their strange behavior.	Yes	Girls barked like dogs. (page 46) Girls danced strangely in the woods. (page 46) Girls fell to the floor and screamed. (page 46)
The Puritans used information from a person's past to charge them with witchcraft.	No	Puritans used five types of evidence: (1) recite the Lord's Prayer; (2) physical marks, like moles; (3) testimony; (4) ghosts; and (5) confessions. (page 48)
The cause of the witch scare in Salem was never identified.	Yes	No one knows the truth. (page 49) Stress from attacks on the colony, smallpox, and other things created fear and a need for blame. (page 49) Historians make educated guesses. (page 49)

Figure 4.1: Find evidence to support claims.

We recommend that teams limit the number of claims students explore during
a reading assignment. Three to five high-quality claim statements will provide students with plenty to think about during their reading. The claims should connect
to big ideas that the class will be discussing throughout the entire text, not something that is relevant to only one small section. For example, the reading in this
example about the Puritans and Salem may be part of a larger unit on problems

that the Puritans faced while adapting to life in North America. In this way, students have to think deeply about the claims.

Adaptations

When first introducing this strategy to students who are struggling with the literacy aspect of the strategy or other students who struggle with reading, it may be helpful to provide one claim statement at a time. Students may need more time to process each claim and would likely benefit from additional supports when identifying evidence. This may require additional modeling from the teacher, showing students how they would identify evidence that supports a given claim. As students become more skillful with identifying accurate evidence, the teacher can remove some of those scaffolds and encourage students to work more independently.

It may also be important to ensure students learning English are familiar with the academic vocabulary of *claim* and *evidence*. Additionally, there may be opportunities to discuss how different cultural norms influence some of the claims the lesson explores. Do the people or societies featured in the text respond differently than those from the student's primary culture? What influences those differences?

To raise the rigor for proficient readers, encourage them to write their own claim statements as a prereading activity, and identify evidence that supports or counters their own thinking. Given that the original activity provides students with claim statements, this adaptation takes away that support and encourages students to independently analyze the big ideas of a text.

Text-Dependent Questioning

When social studies teachers show their students how to read and question a text, they provide a clear model for students to think like a social studies expert. We often tell the teachers we work with that good readers may not have all the answers, but they do have the right questions. Those questions are an important step on the road to find valid and reliable answers, an essential 21st century skill. However, for many students, questioning does not come naturally. They often view reading as a one-way street where the author provides information and the reader absorbs it. This mentality creates a passive reading experience. Buehl (2017) calls this *pseudoreading*. Pseudoreaders skim texts, process information at only a surface level, and usually read to get it done rather than for comprehension or understanding. They are the opposite of critical readers.

To push students to be more active and engaged readers, consider how your team's teachers can demonstrate that reading should be a dialogue between the text and the reader. By prompting readers to generate questions, the text-dependent questioning strategy allows students to become more active, metacognitive, and thoughtful during the act of reading. This is a skill that will serve them well throughout their schooling and in life after school.

How to Use

Questioning is a complex skill that has applications across the curriculum. Because questioning skills don't always come naturally for students, they will benefit from seeing a teacher model the ways an expert generates questions while reading. Consider beginning a reading assignment by taking five minutes to model the questioning process for students using the tools we provide in this section. Be mindful of the nature and complexity of questions that you want your students to generate, and consider providing a bookmark or graphic organizer with a list of question types or examples. Some teachers have found Bloom's taxonomy revised (Anderson & Krathwohl, 2001) to be a valuable questioning hierarchy. This hierarchy establishes a thinking progression across six levels from low-level to high-level thinking: (1) remembering, (2) understanding, (3) applying, (4) analyzing, (5) evaluating, and (6) creating. (Learn more about this taxonomy in the Importance of Postreading Instruction section, page 108.) Depending on your students' reading levels, your team may find it helpful to simplify these taxonomy labels to make them more student friendly for questioning purposes.

When determining appropriate types or categories of questions you want students to focus on, be mindful of the purpose of the task and use your professional discretion to think about the labels that are most appropriate for the students in your classroom. Also, when introducing questioning to your students, it may be more impactful to focus on one or two specific question types. Finding a balance between required or recommended question types and student choice in questioning allows students to develop specific skills with asking different types of questions, and to work at their own level. For example, some teachers may begin by asking students only to practice literal, basic questions at the lowest level of Bloom's taxonomy revised. Later, they may layer in higher-level questions to help students gradually acquire skills at a reasonable pace. We encourage teams and teachers to adapt this strategy to meet the needs of the students in each classroom.

Figure 4.2 (page 92) features a text-dependent questioning graphic organizer that students can use to see examples of different types of generic questions and record questions of their own. (See page 183 for the reproducible "Text-Dependent Questioning Graphic Organizer.") The tool is organized to categorize different types of questions based on complexity, starting with the most basic questions

(*Said what?*) to the most complex questions (*So what?* and *Now what?*). The middle column provides a series of general questions that students can apply to any text. We designed these to inspire students to develop their own text-specific questions that are unique to that text. As students read, they can record their own questions to reflect their thinking while reading.

Social Studies Text-Dependent Questioning for "Sitting Down to Take a Stand" by Sam Roberts and Joe Bubar (2020)		
Question Category	**Questioning the Author or Document General Questions**	**My Focused Questions**
Said what? What is the author or document saying?	What is the author or document telling you? What does the author or document say that you need to clarify? What can you do to clarify what the author or document says? What does the author or document assume you already know?	What does it mean to take a stand? What are civil rights? What or where is the Jim Crow South? What is the story of the Greensboro Four?
Did what? What did the author or document do?	How does the author or document tell you the information? Why is the author or document telling (or showing) you about this event, statistic, era, example, or visual? What does the vocabulary reveal about the author, content, or document? How does the author or document signal what is most important? How does the author construct his or her text or develop his or her ideas?	Why are the Greensboro Four important? What is a sit-in, and how can it be impactful? What were important moments in civil rights history?
So what? What might the author or document mean?	What does the author or document want you to understand? Why is the author or document telling you this? Does the author or document explain why something is so? What point is the author or document making here? What is the author's or document's purpose, and what support (evidence or reasoning) does the author or document present?	How did the Greensboro Four inspire a movement? How can people make an impact on our world? How has the civil rights movement evolved over time?

Now what? How does this connect or apply to what I know?	How does this change my understanding of protesting?
What can you do with your understanding of the author or document?	How does what the author or document says influence or change my thinking? What implications can I draw from what the author or document has told me?	How does this affect my view of current efforts to promote civil rights?

Source: Adapted from Buehl, 2017.

Figure 4.2: Text-dependent questioning graphic organizer.

In addition to (or instead of) using a larger graphic organizer such as this, consider providing a bookmark printed with a list of question types or examples. Consider the text you want students to use with this strategy and choose questions or adapt them from those in figure 4.2. Students will benefit from repeated practice with specific types of questions, but finding a balance between teacher-required or recommended question types and student choice in questioning allows them to establish some control over their approach to learning, focus on developing specific questioning skills, and work at their own level.

The questions students generate from a during-reading activity can evolve into a postreading activity by simply writing answers to their own questions; trading questions with others and discussing answers, providing teachers with a bank of questions to assess student reading; or a combination of these activities. For longer readings, they can also bring the questions they generate to small-group or whole-class discussions throughout the reading, giving them more control in these discussions. Traditionally, teachers often rely on student responses to teacher-generated questions to measure comprehension; however, we often tell teachers they can learn a great deal about how students think and learn based on the questions students themselves generate. For example, if a student consistently miscategorizes their questions, such as labeling a factual detail as a higher-order thematic question, that is an indicator that he or she struggles with that level of comprehension. This can provide formative data for teacher teams as they determine where students need more support with questioning. Overall, this strategy makes questioning skills explicit, and it works with almost any reading task.

Adaptations

Some students may be more equipped to ask higher-level questions than others, and that's OK. Students who have mastered the more literal level of questioning may be ready to challenge themselves with more inferential levels of questioning, while other students may need repeated practice with basic question types. Each student can work at the level where they are appropriately challenged.

For students who qualify for special education, it may be helpful to work on one level of questioning at a time to give them room to practice and become proficient. While working, limiting students to one or two questions in each section can narrow the focus for the students, so it may be appropriate to reduce the number of questions on the form. Narrowing the questions makes the task less overwhelming and the outcome you are trying to achieve more explicit.

Both special education students and students learning English will benefit from modeling. By demonstrating how an expert can generate a specific type of question from a text, students will gain a clearer understanding of the question-generating process. Many times, students don't understand the limits of their knowledge, or they lack the ability to express their knowledge using an unfamiliar language. Modeling identifies and clearly states the strategy's expectations. The more examples and modeling students experience, the clearer the awareness of expectations, which leads to fewer misinterpretations of a task due to learning differences or communication barriers.

Encourage students who have previously demonstrated mastery with questioning to generate fewer literal questions and instead move up Bloom's taxonomy levels to generate increasingly higher-order questions, possibly connecting the ideas in one text to previous readings or content or to their own original writing.

Primary Source Graphic Organizers

The C3 Framework for Social Studies and the CCSS ELA challenge teams to develop a curriculum that teaches students to develop critical-reading and critical-thinking skills. For students to analyze a variety of sources and make historical arguments, they need access to strategies that make meaning of a wide variety of primary source materials. For this purpose, we designed a primary source graphic organizer that is general enough to work with all types of primary sources but focused so students can develop thinking habits that help them process primary source material as a social studies expert. While the organizer appears in the during-reading section of this book, it also has steps for pre- and postreading.

How to Use

Teachers need to model for students how to use the graphic organizer, as illustrated in figure 4.3. (See page 185 for the reproducible "Primary Source Graphic Organizer.") We encourage teachers to adapt the sections to fit the purpose of the reading and the specific course. Using a short primary source, teachers can model the different steps for before, during, and after reading, showing how a historian or social studies expert deconstructs a primary source. After modeling, provide students with multiple opportunities to work with the graphic organizer. When students have repeated practice with the graphic organizer, they have the opportunity to develop reading and thinking habits that will help them to view the primary sources through a critical lens. This approach works best when students use the notes they enter into the graphic organizer for a larger purpose, such as a classroom debate or a historical argument for which students gather evidence from multiple sources.

Prereading	
Type of Document Is this a letter, an article, an advertisement, a government document, or something else?	**Author of Document** Who wrote the document? What prior knowledge do you have about the author?
Date of Document When was the document created? What connects the document to other events or people during this time?	**Source of Document** In what city, town, or country was the document created? Where was it originally published? Do you have prior knowledge about the source?

During Reading	
Style of Document What is unusual about the language used in the document?	**Point of View of Document** Is the document written in the first person, using the pronoun I? How does the point of view affect the reader?

Figure 4.3: Primary source graphic organizer.

continued ⟶

Vocabulary of Document

What are key vocabulary words in the document? What words are challenging to understand?

Postreading	
Main Idea of Document	**Impact of Document**
What is the main point the writer is presenting? Summarize the entire document in two or three sentences with a focus on the main ideas.	What feelings does the document bring up in you? How does it connect to our unit of study?
Questions Raised by Document	**Further Research**
What do you want to know more about? What is still unknown?	What are some other documents worth finding and reading?

When working with the students on this strategy for the first time, walk them through it step by step and explain how each box benefits their reading of the text and future use of the strategy. Encourage students to read and reread the directions within each of the boxes to confirm that they have noted down all the information called for.

Adaptations

At first, students who are struggling with the literacy aspect of this strategy may need additional structure as they work to organize and put down their thoughts. To offer additional guidance and support, for example, provide a specific number of bullet points or partially completed sections. Eventually, you want students to do this independently, but initially providing the extra structure can help narrow the focus and build the confidence students need to continue the reading and not give up. In addition, teachers can create an added space on the graphic organizer for students to provide a general opinion or thought about the reading. This opinion or thought may be important for a future conversation, give students some sense of empowerment, and help provide structure for coming to the correct answer.

For students learning English, simplifying the language or increasing the specificity of column headings on a T-chart or as part of the graphic organizer can help them understand what is important. While some students may be fine with a column that asks them to react to a quote they select from the passage, other students may need more guidance with specific questions, such as *How did this passage make you feel?* or *What do you think will happen as a result of the passage?*

A proficient notetaker is better equipped to articulate what is and is not working well for their understanding of primary sources. Because proficient notetakers will be more successful at appropriately implementing their notes or questions, allow such students more flexibility in the structure of their notes. This allows them to ask the appropriate questions and mark questions within their notes. Note that this isn't to imply any student shouldn't be able to make modifications, but more proficient readers are more likely to identify where modifications are appropriate.

Text Chunking

Social studies texts, regardless of discipline, are packed with content. Almost every sentence holds vital information students need to comprehend to succeed in the unit of study. For many students, that makes reading and notetaking an overwhelming task. We've heard students ask, "How do I know what's important if everything is important?" It's a fair question. Chunking the text allows students to break the text into manageable parts, giving them chances to take thinking breaks and compartmentalize what they've read. Not only does chunking the text make reading less overwhelming, but it helps students slow down their thought process and make meaning from what they've read.

How to Use

Break a larger reading assignment into smaller sections. The length of the sections may vary based on several factors, including text complexity, student readiness, or the reading tasks. For example, although a high school teacher might use longer chunks for his or her more proficient readers (as compared to middle school), he or she might opt to use shorter chunks of text for a passage with particularly complex vocabulary or syntax. A middle school teacher who typically uses smaller chunks might instead use larger chunks of text if the reading is more accessible to his or her students. The length will vary, so teachers should use their knowledge of the text and their students when chunking. In each section, students stop and take thinking breaks—usually focusing on one question or skill—that allow them to either respond to teacher-generated comprehension questions or generate their own questions and thoughts about what they've just read. This could be as simple as generating a supplementary question guide with numbered paragraphs, or the teacher could provide a notetaking sheet where students record observations or ask questions, perhaps using the text-dependent questioning strategy (see page 90). Students may then use their responses or notes to share their thinking with their peers.

Adaptations

Students who struggle with reading comprehension will benefit from taking more frequent breaks to process information. For this reason, chunking a text is a natural teaching strategy for these students. To provide even more scaffolding for the students using this strategy, mark in the text where you would like students to take reading breaks, and consider providing space on the actual text for students to record their thoughts. This may take the form of an empty box added to the text to allow for writing space. You can add a question or assign a task at this break for the students to respond with their thoughts. This is similar to using RATA strategy breaks (see page 86). This concrete structure can become part of the classroom culture, and you can structure all readings in this fashion. As struggling readers move toward proficiency, teachers may extend the gaps between thinking points as students approach grade-level comprehension.

For students who exhibit proficiency, the teacher might lengthen the chunks and provide tasks, such as summarizing or asking questions after each section read.

Lateral Reading

One of the great challenges social studies teachers face is how to support students navigating a limitless sea of online content. Noncredible information sources are seemingly everywhere, and the ways in which consumers of information evaluate and validate source material is evolving. Throughout much of 2000–2010, simply checking the About Us page of a website (for an author's or organization's credentials) or noting the domain (.org, .edu, .gov) might have been enough to evaluate online content's authenticity or expertise. However, there is an ever-increasing number of websites, apps, and other media designed specifically to disseminate false information while appearing credible. Such resources often flash logos and domain names (the part proceeding .com or .org) that appear to be from legitimate individuals or organizations but are, in fact, faked and designed to spread propaganda and misinformation.

Wineburg & McGrew (2018) make the argument that students need to develop lateral (or horizontal) reading skills when evaluating source material, and further contend that it is essential that students develop *civic online reasoning*, the ability to "effectively search for, evaluate, and verify social and political information online." When evaluating an online source, *lateral reading* encourages readers to open a variety of web-browser tabs to read horizontally for the purpose of evaluating the credibility of source material. Readers populate these additional tabs with content related to the subject article *without* exploring contents within the same website. When reading laterally, a reader searches the author or the organization to learn more about their background. Once the reader has found more about the creator of the original content or the authenticity of the online resource, he or she opens additional browser tabs to fact check different claims made throughout the reading. By moving horizontally, both across browser tabs and across a variety of resources, verifying information in real time, the reader adopts lateral reading habits necessary to verify the source and various claims provided in a text.

In a Stanford study (Wineburg & McGrew, 2017), researchers found that even well-respected historians struggled to determine the authenticity of several websites, often relying on outdated and time-consuming authentication methods such as the CRAAP test (currency, relevance, authority, accuracy, and purpose; CRAAP Test, n.d.) or a variety of authenticity-focused checklists. The study revealed that when teachers and students employed lateral reading to validate a web source, they were much more successful in evaluating the credibility of the information.

How to Use

If students are to successfully evaluate and cite valid and reliable information from digital sources, teachers have an obligation to teach students strategies to evaluate a source's credibility and motives. Modeling the process is necessary for students to see how this works, and learning to do so may provide an opportunity for teacher teams to collaborate with the school librarian or research experts. In fact, our own librarians introduced us to the idea of lateral reading by building a lesson around the website Minimum Wage (www.minimumwage.com). They showed our students—and our teacher team—how a website can be set up to appear legitimate, including a seemingly credible, although ambiguous About Us page. However, by conducting lateral reading about the organization behind the website, the Employment Policies Institute (EPI), it soon became clear to students

(and teachers) there was more there than met the eye. First, the librarians con-
ducted a search on a projected screen, typing *Employment Policies Institute*. Once
again, it produced another website that appeared reputable. Further searching
brought up the EPI Wikipedia page, which revealed that the group is funded by
a lobbyist for the restaurant, hotel, and tobacco industries. A return to the search
brought up the site Source Watch (http://sourcewatch.org), which also identified
the EPI as a front group for the lobbyist. This helped the students to see that if they
are to use a site they find while searching, they should be aware of its hidden biases
and agendas. After seeing this approach modeled for them, teachers asked students
to conduct their own lateral reading by using search engines and responding to
specific questions, such as the following.

▸ What did you search for when conducting lateral reading?

▸ What websites did you visit during your lateral search?

▸ What, if any, biases may impact your reading of your source?

Adaptations

When implementing lateral reading with students who struggle engaging with
online content, including students in special education or students who are learning English, it's beneficial to model exactly where to click or how to conduct a
search when evaluating a horizontal source. Also, giving students a specific number
of links to click or keywords to search will help them not to feel overwhelmed
when clicking off the original webpage. Also, providing students with a checklist
of what to look for when evaluating the webpages they encounter gives them useful
criteria to support their evaluation.

Have proficient students write brief justifications for what makes an author or
source valid and reliable. While time consuming for some students, adding this
layer of high-level thinking to their lateral reading encourages students to be even
more critical when evaluating resources.

Build Reading Stamina

Just as it requires increasing stamina to run a mile, five miles, and then ten miles,
developing readers need to build students' stamina for focusing on a text. Building
reading stamina is especially important as students begin engaging with longer and
more demanding historical texts. Students who don't have the ability or stamina to
make it through such texts need coaching so they can learn to do so. As they increase

their stamina, students heighten their ability to consume an entire text with the understanding necessary to grasp historical and sociological events and concepts.

How to Use

This strategy can be very similar to text chunking (page 97), but instead of building stamina through chunked excerpts or shorter texts, this strategy focuses on consuming whole texts of increasing length. To facilitate this, explain to students how they can vary the way they read while consuming a text. They should read to themselves, read to someone else, and have someone else read the text to them. Teach students to set a goal for the amount of time they spend reading and increase this goal regularly. Early in the school year, this might be shorter lengths of time (ten to fifteen minutes). Then, increase the time as the year progresses. Middle and high school teachers should be mindful of students' homework loads, so capping the reading time to around thirty minutes is a good ceiling. Ensure students understand that reading quickly but without understanding is the opposite of this goal. The more they read, the better their reading will become, and their stamina will increase steadily.

This strategy can be very effective when the teacher models the different types of readings in the class and explicitly shows students how, each time they increase their goal for the amount of text they want read, it helps improve their overall reading stamina. As students increase their stamina, celebrate their progress by having them chart their improvements so they can visually see their growth. Again, the purpose isn't to encourage speed reading but to honor building stamina and improvement in learning.

Adaptations

One way to adapt this strategy or to visually show students' progress is to track a student's work over time. The students can track either how many minutes they read or the number of pages read, and then they can see how those numbers increase or decrease based on what they are reading. Although speed reading is not the goal, students who struggle with reading or who are learning English can expect the volume of content they consume during a given period to increase as well. As students develop, encourage them to increase or decrease their reading speed depending on their prior knowledge of the content they are reading. Students can increase their reading rate with content they are familiar with and proficient in while deliberately slowing their pace when they are not familiar with other aspects of the content. This also helps them to monitor their learning and build their overall stamina.

For proficient readers, challenge them to spend more time daily on focused reading outside of class. Classroom hours are limited, but time at home can help students expand their stamina and better prepare them for the rigors of lengthy collegiate-level reading.

Considerations When Students Struggle

Students who struggle with during-reading exercises will benefit from repetition (reteaching) and exposure to multiple reading strategies that engage their higher-level-thinking skills. As with prereading, struggling during reading can occur for a variety of reasons: the text itself may be too high level and create frustration; the text, as a primary source, may use archaic language unfamiliar to most students; some students may need additional instruction and practice with a specific strategy; or the vocabulary or literacy concepts involved may be too challenging. Every student is bound to encounter their own unique challenges. Here are some additional questions to consider when you encounter students who are having a hard time applying during-reading strategies.

▸ **How might you help students improve the focus of their annotations and notes?** A few ideas to help students with notetaking and annotating include modeling, focused practice in small groups for more individual attention and discussion, and providing codes and symbols to make annotations go faster (the time-consuming nature of annotation can often turn students off from even practicing). For example, have students star passages near important dates to remember, write a question mark near a confusing passage, or write a *D* near important details. You may note we repeatedly tout the importance of modeling for students. When students have good models, they can better emulate the skill.

▸ **How might you help students with shorthand to speed up their notetaking?** As noted for annotation, the first key to helping students with shorthand is to provide codes for important concepts they need to address. Teachers can also customize shorthand approaches on a text-by-text basis, increasing the sophistication of students' notetaking as they progress. Teachers should constantly model shorthand for students who struggle to cut down notes to essential knowledge. For example, teach students specific notetaking symbols they can use while reading. As students consistently use the symbols throughout the year, their confidence in applying them will grow and benefit them in other courses and subsequent grades.

▸ **How might you help students who have vocabulary deficits and difficulties?** Provide students who need extra support with vocabulary word lists and encourage them to look up words before engaging in a reading. It is important to also teach students how to monitor their understandings of vocabulary by pausing when they see unfamiliar words and not only take the time to look them up but practice using them.

▸ **How might you help students stay focused during a reading?** Simply using the strategies in this chapter will support struggling students with focus during their reading. When students actively ask questions, take notes, and engage in other active reading strategies, they develop their ability to stay focused. However, teachers also need to help students recognize when they are distracted and patiently provide strategies for re-engaging with the text. Some of these helpful fix-it hints include asking a question to help students refocus or encouraging them to go back a paragraph or page and skim to see if they can identify where meaning broke down (where they "spaced out" or otherwise lost focus). Such interventions enable students to re-engage with the text at that point. Through working on stamina and other strategies, students should grow with their ability to stay on task.

We believe that the strategies in this chapter provide practical instructional approaches for collaborative teams to turn to when trying to increase student engagement while reading. At the same time, teachers should solicit feedback (gather data) from students to find out what is working best and where they find the strategies to be less helpful. This sends the message that teachers value students' feedback, and seeking students' input provides them with a sense of ownership over their learning. Teachers can report that feedback to their collaborative team, and the team can use that information along with other formative assessment data to make informed decisions about where to take instruction in the future. The response may be to adapt an existing strategy to use in specific cases, or it could be to set it aside in favor of a different strategy that may have a greater impact on students' learning. It's important to remember that students need practice with strategies—it often takes teachers (in terms of instruction) and students (in terms of learning) multiple repetitions before they can master a strategy.

thinking
BREAK

Review the strategies in this chapter. How might you
help students who struggle to make progress?

Considerations When Students Are Proficient

For students who have demonstrated proficiency, teams should collaborate around adopting instructional approaches and strategies that are challenging and thought-provoking. In addition to the adaptations in this chapter's strategies, here are a few considerations for further differentiating instruction for proficient readers.

▸ When grouping students so that proficient readers are working with other proficient readers, encourage them to share their thinking with their group while reading. This challenges each student in the group to increase his or her own reading-comprehension toolbox.

▸ When varying groups to match proficient readers with non-proficient readers, encourage proficient readers to share their process for making meaning from texts. This allows them to serve as mentors and leaders to other students in the classroom. Consider revisiting the Read Aloud, Think Aloud strategy (page 86). Using this strategy creates metacognitive space for students of mixed reading abilities to share their expertise or ask questions about how to construct meaning.

▸ Develop reading tasks that are appropriately challenging. Many proficient readers are risk averse. Pushing these students outside their comfort zones doesn't mean creating more work for them; rather, it ensures the time they spend challenges them. Having a bank of media texts available is one approach to allow students to continue their learning if they finish reading a text faster than some of their peers.

▸ When you embed into your during-reading strategies choices for students, ensure those choices are available to all students. The implication should never be that only proficient readers are entitled to choose. When students of all proficiency levels have choices, it sends a message to all students that they have a say in their learning.

During reading, teachers commonly hyperfocus on ensuring still developing students attain the minimum learning necessary for success on assessments and other postreading tasks. However, a school that functions as a PLC doesn't just want all students to learn; it wants all students to learn at *high* levels. Encourage proficient readers to stretch and break out of their comfort zone. By peer mentoring, students reinforce their ideas and gain important confidence in working with others. Also, by providing times for all students to read supplemental materials beyond the text—for interest and learning—proficient readers can move beyond the general learning targets while their peers choose texts that reinforce current targets. Ultimately, the more thinking teachers can encourage proficient students to undertake, the more they will solidify their knowledge and remain engaged in learning.

Wrapping Up

Thoughtfully planning during-reading activities allows students to practice comprehension and engagement. If you provide them with ample opportunities to practice and repeat strategies, students will internalize those strategies, ensuring they are strategic readers in future years and courses. During-reading proficiency does not happen overnight, and it may not even happen over a month or year. Developing the ability to read texts with consistency requires a lot of modeling and practice. Without frequent use of during-reading strategies, like an unused muscle that atrophies over time, reading skills will also stagnate or even decline. To ensure students are proficient and college ready, engage them in important during-reading strategies as historians exploring social studies content.

Collaborative Considerations *for* Teams

- Which strategies are best suited for your students? Consider the following.

  - Do student data indicate specific strategies that would benefit differentiated groups of students?

  - How can the team supply background information and a reading purpose that engages students and helps them focus on the text?

- Which strategies in this chapter are best suited for a particular text and desired outcomes?

- Which strategies in this chapter are best suited for the targeted C3 Framework for Social Studies and CCSS ELA?

CHAPTER 5

Postreading Strategies

One of the best aspects of our collaboration with social studies teachers is, whether or not they realize it, they often use literacy ideas in postreading, even if those ideas aren't necessarily a deliberate part of their everyday instructional practice. For many social studies teachers, there is little distinction between postreading and reading assessment. For decades, teachers have assessed what students have learned from texts via multiple-choice exercises, short-answer questions, and essays of all kinds.

Ultimately, our collaborations reveal that there are possibilities for teachers to instruct students on what to do after they finish reading to help solidify their learning. Encouraging students to continue thinking like a historian or social studies expert as a postreading strategy can help guide expectations for their learning before engaging with a formal assessment. When teachers ask students to actually use and apply the literacy skills they are working on—not only to obtain disciplinary information but also to apply and share that information—the thinking and knowledge they produce can be impressive.

Students can and should use postreading literacy strategies not only to analyze their own thinking but also to share and converse with others, thereby developing critical lenses and thinking skills that will benefit them as literate, college-ready consumers and producers of written content in the areas of history, psychology, sociology, economics, and so on. When this occurs with fidelity, we find students are increasingly likely to be more successful with formative and summative assessments within these disciplines.

This chapter explores how you can guide students toward deeper understanding *after* engaging with a text. We begin by providing a rationale for postreading instruction, including how it fits within the C3 Framework for Social Studies and the CCSS ELA. We then offer concrete postreading strategies your team can adopt to help students make meaning out of rigorous texts. The chapter concludes with considerations for addressing students who continue to struggle and for those who show mastery.

thinking
BREAK

What types of postreading activities do you typically use in your classroom?

- ○ Do you have a pattern of activities that you use with students after they complete a reading?
- ○ Are they always focused on content?
- ○ Do they require any literacy skills?

The Importance of Postreading Instruction

A critical aspect of effectively using postreading strategies in your classroom is to engage students' thinking at higher cognitive levels. To that end, we find that Bloom's taxonomy revised (Anderson & Krathwohl, 2001), which we introduced for the text-dependent questioning strategy (page 90), is one of the most effective ways to illustrate how students build on their thinking from basic (low-level) to more complex (higher-level) thinking. Figure 5.1 illustrates the thinking levels in this taxonomy.

Notice that remembering and understanding are the basic building blocks of the taxonomy. Figure 5.1 looks like an inverted triangle because remembering and understanding are the foundation of thinking, but they are basic skills and thus smaller. The levels expand as they move up due to the more complex level of thinking required, represented by a larger level that creates the inverted triangle. A lot of the strategies and work readers do in prereading and during reading are geared toward gathering information (remembering) and comprehending (understanding) it.

The reason the building blocks are important is that without them, none of the higher-order-thinking skills are possible. For example, remembering key events, people, and places is a foundational skill for history that students must build from. To understand how cause-and-effect relationships interact throughout different historical periods, students must already have a foundational knowledge to build on. Only once students have an understanding of information can they begin to think more critically about it by engaging in skills that work at the apply, analyze,

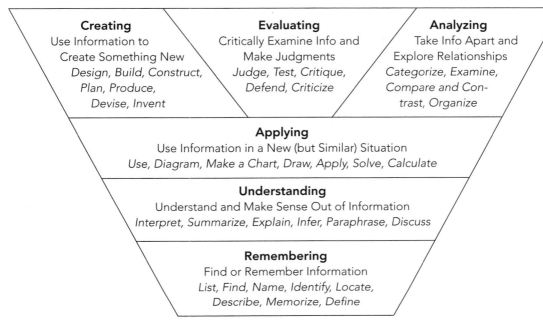

Figure 5.1: Thinking levels in Bloom's taxonomy revised (Anderson & Krathwohl, 2001).

evaluate, and create levels (Anderson & Krathwohl, 2001). Teachers need students to work at these higher thinking levels during postreading activities designed to engage and challenge students.

Collaboration Around Postreading Activities

As with prereading and during reading, it is important for collaborative teams to remember the four critical questions of a PLC when making instructional decisions: (1) What do we want students to learn? (2) How do we know if they've learned it? (3) What do we do if they haven't learned it? (4) What do we do when they have learned it? (DuFour et al., 2016). However, of particular importance at this point is the second question: How do we know students have learned the material? It is through deploying formative postreading strategies that both teachers and students can gain the most understanding of what students do or do not know *before* they take an assessment. In particular, these strategies produce the data necessary for teacher teams to collaborate and make decisions about what and when to remediate and when it is time to move forward to new learning targets and skills. Similarly, these activities help students to self-assess their own levels of remembering and understanding to engage their own learn-to-learn skills (the ability for students to teach themselves).

For example, one of our AP economics teachers reached out to her literacy coach when she knew that students were struggling to make sense of their college-level textbook. She had engaged in pre- and during-reading instruction, but students struggled to make connections between different economic concepts within a specific text chapter. Given the reading was dense and complex, we got to work on ways to support students' abilities to synthesize their reading and build bridges between different sections of the reading. We ultimately settled on two different graphic organizers designed to help students make connections between economic concepts. This allowed students a choice for how they would synthesize their newly acquired content knowledge after reading. Students then self-reported how much the strategy helped them on their assessment, and they were able to continue using the tool they selected or try the other option. This collaboration between teacher and literacy coach ultimately helped provide answers to all four critical questions.

C3 Framework for Social Studies Connections

An examination of the C3 Framework for Social Studies and CCSS ELA reveals the necessity of assessing student reading skills along with students' ability to access and use the content of their history, civics, economics, and other social studies texts. Deeply diving back into the CCSS ELA, the necessity for postreading activities dependent on remembering and understanding is clear. Basic summarizing and drawing conclusions rely on a solid foundation of basic comprehension. Beyond the foundational postreading needs of understanding and remembering, students must also assess content, evaluate arguments, synthesize sources, and support their own thinking, all of which are skills that students can develop and practice with varied and comprehensive postreading activities and exercises.

As we discussed in the previous chapter, there is a direct connection between the four dimensions in the C3 Framework for Social Studies and CCSS ELA skills. Reviewing the CCSS ELA, in particular, highlights the necessity for postreading activities dependent on remembering and understanding. So, as we did in chapters 3 and 4, table 5.1 provides potential connections between the strategies in this chapter and the postreading skills in the CCSS Anchor Standards for Reading, and table 5.2 does the same using the four dimensions in the C3 Framework for Social Studies. See the C3 Framework for Social Studies Connections section in chapter 3 (page 45) for a detailed list of the Anchor Standards for Reading.

Table 5.1: CCSS ELA Postreading Strategy Connections

Postreading Strategy	R.CCR.1	R.CCR.2	R.CCR.3	R.CCR.4	R.CCR.5	R.CCR.6	R.CCR.7	R.CCR.8	R.CCR.9	R.CCR.10
Five Words	✓	✓		✓						✓
Student-Generated Questioning Taxonomies	✓	✓	✓	✓	✓	✓				✓
Synthesize Sources and Connect to Prior Knowledge	✓	✓	✓			✓	✓	✓	✓	✓
Formative Quick Checks Using Online Quizzes	✓	✓	✓	✓	✓	✓		✓	✓	✓
What Does It Say? What Does It Not Say? How Does It Say It?	✓	✓	✓	✓	✓	✓				✓
Targeted Entrance and Exit Slips		✓	✓		✓	✓				✓
3–2–1	✓	✓	✓						✓	

Source for standards: NGA & CCSSO, 2010.

Table 5.2: C3 Framework for Social Studies Postreading Strategy Connections

Postreading Strategy	Dimension 1: Developing questions and planning inquiries (R1)	Dimension 2: Applying disciplinary concepts and tools (R1–10)	Dimension 3: Evaluating sources and using evidence (R1–10)	Dimension 4: Communicating conclusions and taking informed action (R1)
Five Words	✓	✓	✓	
Student-Generated Questioning Taxonomies	✓	✓	✓	
Synthesize Sources and Connect to Prior Knowledge	✓	✓	✓	

continued →

Formative Quick Checks Using Online Quizzes		✓	✓	
What Does It Say? What Does It Not Say? How Does It Say It?	✓	✓	✓	
Targeted Entrance and Exit Slips	✓	✓	✓	
3–2–1	✓	✓	✓	

Source for standards: National Council for the Social Studies, 2017.

Strategies for Supporting Students in Postreading

Before we introduce our approach to postreading strategies, it is important to understand that there is significant overlap between what one might consider a postreading strategy and an assessment. In fact, it is accurate to say that you could use many of the strategies in this chapter for formative or summative assessment purposes, depending on the team's learning targets and goals. In chapter 7 (page 157), we discuss our ideas for how strategies like those we list in this chapter can serve as assessments, but in this section, we focus on what makes a strategy effective as a postreading activity.

Because they are quick and easy to create, administer, and grade, a teacher's go-to postreading strategy is often a multiple-choice assessment. However, the problem with multiple-choice assessments is they generally "fail to assess students' productive skills (e.g. writing or speaking) and to prepare students for the real world" (Abosalem, 2016). Multiple-choice style questions tend to make students passive participants on theoretical information and often don't assess the more active, complex real-life skills NGSS and the skills of literacy intend to promote (Abosalem, 2016).

Another issue with multiple-choice assessments as a postreading strategy is that they do not reflect the nature of information and how people use it today. Because information is available at the touch of a keyboard or through a quick oral query to a smartphone, memorization is not the essential skill it once was. Instead, as new standards in science, mathematics, social studies, and other content areas show, the world demands that students can *use* information (International Society for Technology in Education, n.d.). You can see this reflected in the active verbiage we highlighted in the C3 Framework for Social Studies Connections sections throughout this book. Since your teacher team wants to engage students in using information at higher levels, team members also need to quickly determine whether students can obtain such information and use the information. For this reason, postreading really falls into two categories: (1) quick checks to assess comprehension and information gathering and (2) more in-depth checks to allow students to use and engage with the information at challenging levels.

We have accompanied each of the following postreading strategies with an explanation and provided differentiation options for students who struggle—in particular, students who qualify for special education and students learning English—and students who already show high-level proficiency. As in previous chapters, strategies include information regarding why and how to use them. We hope these strategies will spark ideas for you and your team to use and adapt in your classrooms.

Five Words

It's easy to take for granted that students can prioritize the information found in a text and then summarize it. There are times in our own classrooms when we are taken aback when we ask students to summarize a passage, only to find they lack the ability to do so. Unsurprisingly, if students do not receive instruction that provides them with the tools to summarize a complex text, many will struggle to prioritize the information they've read. This strategy asks students to summarize the reading by participating in a close-reading exercise. Students have the opportunity to identify the most important information, share that information with a small group or the whole class, and construct a summary that will include all of the most important information in the text. Not only does this strategy assist students in determining the most important literal information, it requires that they evaluate the importance of specific words or details in the text that build toward higher comprehension levels.

How to Use

Before students read, ask them to mark the text any time they come across a key detail or main idea. We often encourage students to underline or place an asterisk in the margins. This step is essentially an annotation task for students to complete during reading as preparation for summarizing during the postreading activity. When students are finished reading, they must select five words or phrases (teachers can adjust this number as necessary) among those they underlined or marked with an asterisk that represent the *most* important information. This requires them to evaluate all the information in the text and prioritize what is most critical to understanding. Students can record their individual responses on a form you provide, like the one pictured in figure 5.2 (page 114). (See page 187 for the reproducible "Five Words Recording Sheet.") Tell students that they should be prepared to defend their

Directions: While reading "Reply of the House of Commons to King James I, 1604," underline key words and phrases. After reading, choose the five most important words from the reading, and add them to the Individual Selections column. When instructed, as a small group, discuss the words in the Individual Selections column and come to a consensus on the five most important words that the group agrees on. Add those words to the Group Consensus column.

Individual Selections	Group Consensus
1. Misinformation	1. Privileges
2. Inheritance	2. Withheld
3. Representation	3. Representation
4. Judge	4. Judge
5. Controlment	5. Liberties

Discussion Questions

Which words can your group agree on?

The group settled on two of the words I had on my individual list. After reading the source, we tried to agree on words that captured the most important details for the House of Commons' disagreement with King James. We think they are trying to have independence from the King.

Which words led to disagreements?

I wanted to keep the word "controlment" because they were trying to say they didn't want to be controlled by the king, but we agreed on liberties as a better word because it said what they want, not what they don't want.

How did your thinking change as a result of your discussion?

This was a very hard source to understand because the language is so old. Talking through these words helped me to get the main idea that the House of Commons wanted more control over laws.

Source for reading: Pearcy & Dickson, 1997.

Figure 5.2: Five words recording sheet.

word choices for the group work that comes next. This last step is where higher-order thinking comes into play, as students will inevitably share their thinking behind why some details are more important than others. This requires them to make connections or draw inferences to support their arguments. In this sense, the strategy moves beyond a summary strategy and delves into close, analytical reading.

Next, have students work in small groups of three to five to come to a consensus on a list of five words. Include discussion questions on the response form to guide students in this process. Students can begin by acknowledging the most commonly selected words, but when they have a word that is different from their peers, they should defend their selection, and the group will decide whether it is worthy of inclusion on the list. Is that word essential to understanding the text? Is it perhaps encapsulated in a word the group had already chosen? The conversations about word selection are the most valuable part of this exercise, as students conduct a close reading of the text based on their individual selections. Once the group comes to consensus and records its words, students are ready to share with the class.

Each group should display its list, and you can lead a discussion with the whole class, asking questions such as, "What words did most groups agree were essential? Are there outliers that need to be explained to the rest of the class? How can this information support a summary of a complex reading?" Of course, there are no right answers, but the class will evaluate the selections to determine what is most important. Again, the conversations are the most valuable aspect of this strategy. From here, you can choose to have the class write a more formal summary or leave it at a discussion.

Adaptations

When first introducing this strategy to students who qualify for special education, you may simplify it by having them select only three key words or phrases. Starting smaller may make it easier for them to focus on big ideas. Also, providing the students with the first word or set of words gives them an example to follow and use when they are coming to a group consensus. If you give them the structure for their thinking, it helps extend their learning and processing of the information. Teachers may also provide more guidance in the way they frame the task. For example, students might search for the most important words related to a specific history literacy skill, such as cause and effect.

Students learning English may benefit from reflection questions that allow them to identify vocabulary words that are important to understanding. Students might add words to a list of *New Words* or *Words I Wonder About*. Alternatively, teachers can use the reflection questions to help students make cultural observations or ask questions about the culture (or cultures) or historical background in a text.

For students who have mastered the art of summary, teachers might add a critical-thinking task that requires them to analyze, question, or evaluate the topic in some way. For example, a teacher might extend students' thinking by drawing connections to other documents, historical periods, or units of study. In the example for this section, a teacher might encourage students to extend their thinking by considering how other government bodies throughout history might have responded to power struggles similar to the one between King James I and the House of Commons.

Student-Generated Questioning Taxonomies

In several strategies in this book, we emphasize the value of questioning, be those questions in the form of turning titles and headings into questions for pre-reading (page 71) or using text-dependent questions during reading (page 90). We contend that if students know how to ask the right questions, they are on the road toward reading and thinking like a literary expert. Doug Buehl (2017) writes extensively on discipline-specific self-questioning taxonomies. He emphasizes that every discipline requires students to ask questions in its own unique way, and students need explicit teaching of how to ask different types of questions. Compared to text-dependent questioning, this questioning strategy puts increased emphasis on achieving higher levels of critical thinking by asking students to be the ones to build thoughtful questions. It prompts students to reflect on a text they've read and think about and ask questions that they consider the most important—from what is crucial to remember to what new knowledge or meaning a reading creates.

This strategy tasks students with using Bloom's taxonomy revised (Anderson & Krathwohl, 2001) to generate questions that reflect specific thinking levels of the taxonomy after reading a specific text. By asking questions that align to the six levels of questioning (or a subset of the levels)—(1) remembering, (2) understanding, (3) applying, (4) analyzing, (5) evaluating, and (6) creating—students demonstrate different layers of comprehension. This allows both teachers and students to understand where comprehension breaks down. For example, if a student poses the question, *In what ways did ancient Egyptian citizens pay tribute to pharaohs?*, they are posing a basic question about what they read. It asks about a concept he or she should *remember* from the reading. However, if a student labels that question as *analyzing* or *evaluating*, the teacher can intervene and help the student practice developing higher-level questions. In this way, the questions students generate can tell teachers all they need to know about what a student does or does not understand. Sometimes, this may reveal a misunderstanding of Bloom's taxonomy revised. However, in our experience, it often indicates an inability to generate higher-order questions when reading independently. This can be valuable qualitative data for teams as they help to provide instruction to teach students how to think critically when reading social studies texts.

This ability to formatively assess in postreading has even broader implications for teacher teams working in a PLC. For example, a student that focuses his or her questions on the Nile River's effects on Egyptian agriculture while reading an article predominantly about the ruling classes of that society may indicate a misunderstanding of the main thrust of a text. Teachers and teams can work together to examine different types of student questions to ensure they meet the learning targets and display student understanding. If not, the team may have to find ways to intervene and build in more supports and opportunities to practice.

Some teachers we have worked with tell us that asking students to write six different types of questions can be overwhelming and time-consuming. We find that when teachers focus instruction around two to three specific question types, they can hone in on specific questioning skills appropriate to the students and task. This approach works best when differentiating instruction for students at different proficiency levels with the strategy. The more proficient readers will benefit from working higher in Bloom's taxonomy, while students with less reading proficiency focus on building the lower-level skills necessary to reach those higher levels themselves.

In short, if the goal of the social studies classroom is to train students to read, write, and think like social studies experts, we believe one of the most important things we can teach students is knowing how to generate meaningful questions about nonfiction and fiction texts. Students may or may not be able to fully answer their questions, and that is OK. By bringing these questions to light, students engage in close reading, inquiry, and analysis that create several possibilities for learning. Students can use their questions for discussion, writing, and cooperative learning structures. Like most of the strategies in this book, this strategy is not limited to traditional texts. Consider ways students can practice questioning with videos, photos, paintings, charts, graphs, and other nontraditional texts.

How to Use

Good questioning begins with good modeling. It's important teachers take the pulse of their students to identify where they need focused instruction around different question types. We recommend honing in on two to three question types along the taxonomy at a time. Determine questioning types by the task, the text, and the expectations you set for students. If students are working with a complex text and a task they are less proficient with, consider working with lower levels of the taxonomy. If students are proficient with a concept and have had practice with a task, it may be a good opportunity to begin practicing questioning at higher levels of comprehension. Model your process for generating questions for students, and allow them time to practice. We find it's best when teachers model using the text that students will be reading. The teacher can read the beginning of the text and share their thinking process when generating questions at each thinking level he or she wants students to target. This makes the process and thinking transparent for the students.

Allow students to work independently or in groups to read a text and generate their own questions in each of the assigned thinking levels. Teachers and student peers should provide feedback on the different questions students generate. There are many different ways for teachers to utilize student-generated questions to build and sustain disciplinary learning.

▶ Use student-generated, higher-level questions as writing prompts. Students write paragraph responses to their own questions. Likewise, students respond to their peers' questions.

▶ Use student-generated questions in a small-group discussion. Have groups begin at the bottom of the taxonomy to clarify any misunderstandings of what the text says, and work their way up to analytical or big-idea questions that are more implicit in the text.

▶ Collect and check students' self-generated questions to see how well students understand the various question types. Are they mislabeling a question? Are they unclear about the content? Are students demonstrating they are ready to take on more sophisticated question types? Sometimes the questions themselves tell teachers all they need to know about their students.

Figure 5.3 is an example of how one social studies teacher modeled questions to help students make meaning from a political cartoon in a U.S. history course studying Irish immigration (Stanford History Education Group, n.d.). (See page 188 for the reproducible "Student-Generated Questioning Taxonomy.")

Using figure 5.3 as a model, the teacher—with support from students—asks a variety of questions that help them organize their thinking. Notice how the teacher models all levels of questions from Bloom's taxonomy revised (Anderson & Krathwohl, 2001) to match the academic vocabulary of the classroom. Also, there is a range of question types that allow students to think at both the literal and inferential levels. Literal questions tend to exist within *remembering* and *understanding* levels, and inferential and higher-order questions can occur from the *understanding* level to the top of the taxonomy. From there, the teacher focuses

Directions: After viewing the historical political cartoon depicting the artist's take on Irish immigration, note the important information it conveys and write basic questions that will help you increase your thinking level about its message. Use the following chart for help.

Level of Thinking	Comprehension Self-Assessment	Focusing Questions
Creating	I have created new knowledge about the past.	How does my analysis of this cartoon impact my understanding of immigration in the 19th century? Today?
Evaluating	I can critically examine this author's arguments about the past.	What biases or agendas does the artist or magazine have that would influence the messages in the cartoon? Are there other sources that support or refute the viewpoints expressed in this cartoon?
Analyzing	I can understand why this piece was created by placing it in a historical context.	What historical events were taking place in the 19th century that are reflected in this cartoon?
Applying	I can use my understanding to better comprehend how the past influences my life and world.	How does the depiction of Irish immigrants connect to ways that contemporary immigrant populations are portrayed in today's media?
Understanding	I can understand what the author is telling me about the past.	What is the Irish Declaration of Independence described in the caption? How is the Irish immigrant depicted in the cartoon? What stereotypes are perpetuated in the image?
Remembering	I can recall specific details, information, and ideas from this text.	Who are the two women depicted in the cartoon? When and where was this published?

Figure 5.3: Student self-generated questioning taxonomy.

on a new activity that allows students to generate their own questions at various levels of complexity throughout the unit and progress through the taxonomy at their own speed.

Adaptations

This strategy works best when differentiating instruction. For students in special education or who are otherwise still working through the more basic levels of comprehension with a text, the teacher can start off by having a small group focus solely on the literal levels of the taxonomy. Using the example in figure 5.4, the students might narrow their questions around the key details from the political cartoon that are foundational to higher levels of questioning. From this foundation, students might grow to demonstrate higher questioning levels while exploring subsequent resources for the unit. It is OK to have different students working on different levels of the taxonomy while reading the same text. This gives all students an example of others' thinking and questioning and provides a path for students working at lower thinking levels to reach higher levels.

Student Taxonomy Questioning Practice

Directions: As you read, your job is to note important information from the text and write basic questions that will help you remember and understand what is happening in the cartoon. Use the following chart for help.

Level of Thinking	Comprehension Self-Assessment	Focusing Questions
Understanding	I can understand what the author is telling me about the past.	What is the Irish Declaration of Independence described in the caption? How is the Irish immigrant depicted in the cartoon? What stereotypes are perpetuated in the image?
Remembering	I can recall specific details, information, and ideas from this text.	Who are the two women depicted in the cartoon? When and where was this published?

Figure 5.4: Adapted taxonomy practice tool for struggling students.

Encourage students learning English to use the Remembering section of the practice tool to ask about challenging vocabulary words. Teachers might also include the Applying section to explore how the cultural elements of the text are relevant to students' own cultural backgrounds and identities. It may also allow for historical comparisons to other countries students may originate from.

As students progress in their thinking, provide work that pushes them to increasingly higher thinking levels. Conversely, have already-proficient students focus on working at higher thinking from the beginning.

Synthesize Sources and Connect to Prior Knowledge

When students work with different texts in social studies classes, they have to navigate different modes of information. Depending on the text (primary or secondary), historians often read a variety of texts that need synthesis to support historical arguments. To do this, they need to analyze perspectives (to place texts in the proper historical context) and synthesize from a variety of sources to build a well-supported argument. This strategy provides the explicit instruction many students need on how to pull information from various source materials to demonstrate comprehension of social studies content. This strategy is particularly helpful for students building the requisite skills for success when answering a document-based question (DBQ) on an assessment.

How to Use

When students need to navigate between two or more different pieces of source material, you can provide explicit guidance on how to synthesize the readings. For example, if students are reading multiple primary sources related to the Reformation in a world history class, help them make connections between their content knowledge of the Reformation and its application to the primary sources by providing a prompt, such as *Using evidence from the sources, what were the primary causes of the Protestant Reformation?*

Students then write a well-developed paragraph that takes elements from multiple sources to demonstrate a larger understanding of the historical concepts they should know after reading multiple texts. This works with any two or more texts. We encourage you to also use this strategy with nontraditional texts—video, audio, visual art, photography, political cartoons, and so on. It's important not to assume your students can do this independently. Because you will expect students to synthesize throughout the course, it is instructional time well spent for the expert in the room—the teacher—to spend time modeling for students his or her own thinking process as students synthesize multiple primary sources to support an argument. It can be powerful for students to see and hear a teacher's thought process. Students should then have opportunities to practice independently or in small groups after seeing it modeled by the teacher.

Adaptations

For students in special education, teachers can consider ways to scaffold instruction to simplify synthesizing from different source materials. For example, the teacher might provide a claim statement, and students can search through multiple sources to find evidence to support the claim. Similarly, teachers might provide more guidance as they extrapolate evidence from several sources and then work with the class to collaboratively craft a claim statement.

Students learning English should be able to synthesize just as effectively as other students at the same ability level, but the teacher might be mindful of finding source material that is at an appropriate reading level. The vocabulary should be at a level where students can work independently to sift through the source material and identify appropriate evidence while building toward grade- or course-level vocabulary. If working with current events, resources such as Newsela (www.newsela.com) allow students to vary the text level of articles to match the ability of students.

Proficient readers may be ready to search out their own source material to synthesize alongside teacher-provided materials. Focus instruction on how students navigate research databases or the internet to find relevant evidence in different sources. Students use this information to make connections with other texts.

Formative Quick Checks Using Online Quizzes

When your goal is to make sure students are reading and picking out the most relevant information in a text, it is easy to check their progress with a short and simple formative quick check using online quizzing resources like Kahoot! (https://kahoot.com) or Quizizz (https://quizizz.com). Students and teachers can quickly receive formative (ungraded) feedback about whether their understanding of a text is accurate and relevant. As we wrote previously, the use of multiple-choice feedback can be limiting, but it can be useful in efficiently determining where students may need more teaching around course content.

How to Use

Read the text in advance and create a multiple-choice assessment based on the essential information you and your team feel is important for students to comprehend and retain. Access one of the online quizzing platforms and create the quiz. After students have completed the assigned reading, allow them to use their prereading and during-reading notes on the quiz. If students' notes clearly recorded the essential information, they should perform well on the quiz. A quick reflection after the quiz will provide immediate feedback for the students regarding their ability to read, comprehend, and identify the essential information. Combining the students' scores by requesting their responses to follow-up questions, such as *Did you have too few or too many notes? Which types of reading strategies helped you to be most successful?* allows them to evaluate their reading performance. For teacher teams, the results paired with the student reflections may provide valuable data for future instruction, such as whether there is a need to reteach specific learning standards or targets or to plan future lessons.

Adaptations

One adaptation for students in special education is to have them work with a high-proficiency partner the first time. This allows students to build confidence and work collaboratively to build knowledge. In addition, students can do a quick check-in with a partner over the specifics of their notes. They review each other's notes and add any information they feel is missing prior to starting the online quizzes.

For students learning English, teachers might add quiz questions that excerpt passages from the reading that will help determine how well students are applying English language content standards to their reading. This informal assessment may provide teachers of English learners, in particular, with insights about how to dedicate their instructional time to meet students' needs.

Proficient readers will likely do well on fact-based, literal detail questions, so it's important to include extension questions whenever possible. Multiple-choice questions that rise above the literal level may ask readers to select evidence, make inferences, identify connections between different points in time, or use any other number of higher-order thinking skills specific to a social studies discipline. This will appropriately challenge proficient students, and students still working toward proficiency will still have exposure to the more rigorous questioning they will eventually reach.

What Does It Say? What Does It Not Say? How Does It Say It?

Literacy expert and English teacher Kelly Gallagher (2018) created this strategy to develop close-reading skills in the English classroom. While we have had success applying this strategy in disciplines other than the English classroom, it is an especially perfect fit for social studies postreading activities because it requires students to think about the literal, inferential, and structural elements of a text. It also works very well with small-group or whole-class discussions.

How to Use

After reading a text, have students identify the answers to three questions.

1. **What does the text say?** This refers to what the text says literally. We often tell our students to identify something they can put their finger on, something that the text literally states. What makes it important?

2. **What does the text not say?** This requires students to make an inference about something that is implied in the text. It could also be important information that the text doesn't include but is something the student wants to know more about or feels is relevant to the reading.

3. **How does it say it?** This requires students to explore the structural elements of the text. For example, a student might examine and describe the text features of a historical document.

After students identify one detail for each question, they share their work with the whole class or a small group. Because they have had time to think, all students should be equipped to participate in the discussion, and the class can build understanding as they closely analyze the text through literal, inferential, and structural lenses.

Adaptations

Students in special education might benefit from exploring one question at a time. Teachers can make a determination about which question is most appropriate for the students, the text, and the task. In our experience, struggling readers may benefit from a limited focus to the first two questions; in this case, we often omit the *How does it say it?* question when students are still working on following literal details from a text. Another adaptation to this strategy is to provide students with an example along with each of the three sections. So, as the teacher, you give them an example of either what the text says, what the text doesn't say, or how the text says important information. Giving this extra detail helps structure thinking for struggling readers, narrows their focus, and provides a starting point for their thinking.

For students learning English, the *How does it say it?* question may provide opportunities for students to analyze English grammar and writing conventions in addition to absorbing disciplinary vocabulary and writing. This might be an opportunity for students to explore both disciplinary and modern vocabulary. For historical texts, in particular, if students are struggling with archaic language, it may be helpful for teachers to modify texts with glossaries or footnotes to support understanding.

Encourage proficient readers to extend their thinking by asking higher-order thinking questions that deepen their knowledge and the conversation around the text. (See the taxonomy in Student-Generated Questioning Taxonomies, page 116, for examples of higher-order-thinking questions.)

Targeted Entrance and Exit Slips

When determining whether a student has understood a text, one of the more tried-and-true postreading exercises to use as a formative assessment is an entrance or exit slip. These low-stakes check-ins allow teachers to take inventory on who is mastering the content or skills of a lesson or reading assignment. As part of this strategy, we encourage adding opportunities for students to self-assess so they can see clearly the criteria for how teachers will evaluate them.

How to Use

Provide students with a targeted writing task (as they enter the class or before leaving class) that requires them to articulate an understanding of a lesson or reading task and turn in their written response. For example, after facilitating a reading about the legislative branch of U.S. government, you might ask students to write on a notecard their answer to the prompt, *In your own words, without using notes, explain how a bill becomes a law.* Notice the wording in this prompt. One way to ensure that students have truly mastered the material is to ask them to explain a concept in their own words. This helps avoid the temptation to rewrite or copy directly from the text. You can also encourage students to extend their learning by applying the content to a broader context, such as *How must lawmakers collaborate to help pass legislation into law?*

The possibilities for questioning are endless, and it is a simple yet effective method to take the pulse of the classroom. Figure 5.5 (page 126) provides a useful example of how easy it is to craft an effective entrance slip to distribute to students as they enter the classroom. In this example, before class starts, a middle school teacher wants students to know what they should remember the day after a class reading of the Declaration of Independence.

On the lined side of the notecard, please respond to the following question.

In your own words, explain one or more of the main complaints that led to the Declaration of Independence.

On the unlined side of the notecard, please respond to the following question.

Based on your understanding of the Declaration of Independence, were the framers justified in declaring independence from Britain?

Figure 5.5: Targeted entrance slip example.

Before submitting their answers to the teacher, students can share their responses in collaborative groups. This type of entrance slip provides an opportunity to compare thinking among a small group of peers and helps group members prepare for the lesson or reading that follows. Conversely, using these as exit slips for students to complete after class to return the next day (or submit electronically that afternoon or evening) helps teachers determine common patterns in students' responses prior to the next class meeting. With these data, teachers can adjust instruction accordingly. The entrance or exit slip is a simple yet effective method to take the pulse of the classroom's understanding of a learning goal.

Adaptations

For students in special education, tailor questions to a specific student's needs to assess his or her level of comprehension. This may require the teacher or team to create differentiated slips. If students require significant guidance, teachers could even start a new or unfamiliar topic with multiple-choice questions instead of writing prompts. (Remember that the goal is still to bridge students to grade- and course-level learning.) Teachers can also use these slips for either literal or inferential questioning. Using the example from figure 5.5, the teacher might revise the open-ended task of identifying one or two complaints to include a specific text excerpt from the document for students to analyze and then ask students to underline the key words in the passage that show whether the framers were justified in breaking away from Britain.

For students learning English, teachers can provide specific vocabulary words they want students to use in their response. This provides a structure for their thinking and gives them some prompts to start their writing.

Teachers can extend the learning for proficient students by moving away from more literal questions and providing opportunities for students to explore analytical tasks. For example, a teacher might adapt some version of the five words strategy (page 113) that asks students to identify the five words from the text that best support a historical argument. Other options include asking students to unpack specific words or phrases that connect to other units of study.

3–2–1 Activity

The 3–2–1 activity is similar to the entrance and exit slip strategies in that both allow teachers to receive quick, formative feedback on student thinking following a reading. Through this strategy, students use an organizer to focus their thinking around specific, varied postreading tasks.

Teachers use this quick organizer to assess student understanding of a task by asking students to provide different lists aligned with different reading tasks. There are a number of skills that teachers can assess through this tool: the ability to determine main ideas, recognize literal and implicit key details, understand context, grasp perspective, or analyze text (just to name a few). Like many of the strategies in this book, teachers can adapt this activity based on the text and task or on students' needs. Through repetition, students will become comfortable with the 3–2–1 strategy, and it can become a go-to choice for students in need of a nimble formative assessment. Nimble, in this sense, means that it won't take a significant amount of time for students to complete the assessment, and it won't take a large time commitment from teachers to evaluate the student work.

How to Use

After reading a text or a portion of a text, ask students to complete three increasingly complex tasks. In the example in figure 5.6 (page 128), U.S. history students read primary sources connected to the Pullman Strike in a unit on the Gilded Age. The 3 requires students to identify three key details from a newspaper article from the period. The complexity increases with the 2, which asks students to make two personal connections to the period. In this example, the 1 asks for a student-generated question; encourage students to ask specific types of questions. (See Student-Generated Questioning Taxonomies, page 116, for different levels of questions students might provide. See page 189 for the reproducible "3–2–1 Activity.") After providing the tasks, ask students to record three responses to the first task, two responses to the second task, and one response to the final task. As another example, after completing a reading, a teacher asks students to write down three key terms, two key details, and one connection to another text.

3. Identify three key details from the 1894 newspaper article about the Pullman Strike.

2. Write two connections to how this event fits into your understanding of the Gilded Age.

1. Write one question you are wondering about after reading.

1. Workers were upset about wages, so they went on strike. 2. Forty-two of forty-six committee members were in favor of the strike. 3. The company refused to raise wages.	1. Railroads were one of the biggest industries during this period. 2. During this time, unions were trying to earn better working conditions and hours.	1. Were the workers right to go on strike?

Figure 5.6: 3–2–1 example.

This strategy gives teachers quick formative feedback on what students are thinking about the reading. Teacher teams can then use the data they receive to make plans for future instruction or to adjust instruction to support students in future readings. These qualitative data can be powerful for teams that are determining where to focus their instruction. For example, if the team recognizes that students are struggling to make connections between primary sources and the study topic, they can collaborate to come to consensus on instructional strategies to better support students moving forward. They can do so by incorporating more current events that connect to the historical period of study, helping students to see how the unit of study impacts their lives.

Adaptations

Special education teachers can adapt any of the tasks in the 3–2–1 strategy to suit the needs of the students, teacher, or team. If it's appropriate to focus on more key details or terms from a historical text, teachers could ask students to identify three terms, two details, and one important person or event. Teachers can also differentiate tasks within a single classroom, allowing students to have a choice in how they respond to the text or to challenge themselves with more complex tasks. While some students work on more literal details, encourage those ready for a greater challenge to explore historical perspectives, make connections within and outside periods, or analyze historical arguments. All students can work with the same tool on reading skills that meet them where they are.

For students learning English, teachers may ask students to respond in more visual ways, perhaps giving students the opportunity to draw or visualize elements of the reading. A 3–2–1 organizer could ask students to identify three key details, define two challenging vocabulary words, and provide one drawing of a key scene.

Differentiation for proficient readers would call on them to take more of an analytical approach to the text, asking students to provide in-depth analysis of evidence or details in a text.

Considerations When Students Struggle

Students need to practice postreading strategies to develop their analytical and critical-thinking skills, and there are a number of factors that can lead to difficulty for students in this area. For example, students may struggle with the format of a strategy or in achieving the level of thinking it requires. Here are some additional questions to consider when you encounter students who are having a hard time applying postreading strategies.

▸ **How might you ensure students comprehend the literal information required to think more deeply?** Has the team provided frequent formative assessments to identify whether students understand social studies texts at a literal level, consistently identifying key details, places, events, and people? If students are consistently successful, teams can begin to focus instruction on more sophisticated inferential skills.

▸ **How might you push students at different levels to deepen their analysis?** Just because a student may need support with more basic levels of comprehension doesn't mean that he or she shouldn't have regular opportunities to break outside his or her comfort zone. The C3 Framework for Social Studies and the CCSS ELA mandate that students engage with *all* of the standards. This may require teams to examine and evaluate curricular texts to ensure that they allow students to fully engage in meaningful ways. Are the team's text selections instructionally appropriate for the students?

▸ **How might you meet groups of students at different reading levels?** It can be complicated when there are diverse learners in one classroom. Can the team use some of the adaptations suggested in this chapter to ensure that students are all appropriately challenged? Remember that differentiation doesn't mean the teacher needs to plan three different lessons. Instead, think about ways that students can engage in the same strategies but at different points of entry, such as varying the types of student-generated questions.

▸ **How might you tailor the strategies in this chapter so that students can use their strengths for similar tasks or think in a way that is developmentally appropriate for middle school, high school, or college?** Even the most advanced students benefit from strategic instruction. As the rigor increases, students continue to need support with reading analysis. Given the recursive nature of social studies reading skills, the need to increase complexity and rigor of texts is a constant. Students need ongoing instruction on how to engage challenging texts.

As teams begin to explore these questions, there a number of factors to consider. It's important to remember that postreading instruction and strategies are part of the reading process. Teams should consider the effectiveness of prereading and during-reading instruction. If students struggle in the postreading stage, it may be an opportunity for teams to reflect, revise, or reconsider the instruction for prereading or during reading.

Remember that the strategies we provide in this book are designed to help students read, write, and think like historians and social studies experts. Teams should collaborate around ways to adapt the strategies to meet the specific curriculum's requirements as well as the needs of students. Consider adapting vocabulary and terminology so that they align with what students see in the texts and hear in the classroom.

thinking
BREAK

Review the strategies in this chapter. How might you help students who struggle to make progress?

Considerations When Students Are Proficient

There are a number of team considerations for students who have demonstrated proficiency throughout the postreading process. Teams should collaborate to make sure that all students are challenged appropriately. Here are some considerations for differentiating instruction for proficient readers.

▶ Students can step into a mentoring or coaching role, helping fellow students to craft questions or clarify understanding. Teaching the material to others empowers students, while solidifying and extending their own knowledge.

▶ Teachers can pair students with other proficient students to extend their learning through supplemental texts, nontraditional texts, or research. Self-directed learning challenges students to develop their own inquiry and extensions based on their personal interests and questions.

▶ Proficient students can generate their own assessments or questions to challenge less proficient classmates working to practice and develop their own proficiency. Review these student-generated pieces for appropriateness and accuracy.

Wrapping Up

Moving beyond the literal reading and understanding of a text to a more analytical, inferential reading practice takes work. Students do not always default to deep postreading analysis. Encouraging students to be more thoughtful about the material and the application of their social studies knowledge will benefit them as they move through each level of schooling. Students need to realize that when the act of reading is complete, the thinking does not stop. Postreading instruction and strategies are the best ways to ensure that students continue their thinking at a high level after completing a reading. When teams collaborate around ways to teach students to think, recall, question, or synthesize after reading, they are providing students with valuable tools to continue the conversation between the reader and the text.

Collaborative Considerations *for* Teams

🗩 What postreading strategies would be most beneficial for your team to adopt for use in your classrooms? Consider the following.

– How will the team implement these strategies?

– What adaptations may be necessary for the team, task, or students?

🗩 How can postreading strategies provide formative data for your team to impact future instruction?

CHAPTER 6

Writing Strategies

Teachers of social studies are well aware of the rigorous writing demands they place on their students. One of our first coaching relationships was with a U.S. history teacher who taught both AP and college-prep levels. She was increasingly frustrated with students' inability to write historical arguments. Many of her students were relying heavily on writing structures they learned in their ELA classes. Obviously, those are important skills, but they weren't designed to meet the disciplinary standards of historical argumentation. In other words, proficient writing skills look different depending on the academic discipline. In discussing this with the literacy coach, the teacher realized she was telling students about expectations and requirements for their essays (including a thesis, evidence, explanations, and so on), but there was little to no instruction around *how* to write these. Another key takeaway was that the academic vocabulary used in the social studies department differed from what the rest of the school was using to teach writing. For example, the social studies department still used *thesis*, while the rest of the school was saying *claim*. This may seem insignificant, but our literacy team had already seen how writing instruction improved in other disciplines in our school when teachers used a common writing vocabulary.

To address these takeaways, we focused our literacy coaching on scaffolding writing instruction around specific writing skills. The teacher and coach prioritized different writing skills and co-created lessons spread out over the first semester that addressed the various mastery targets included on the writing rubric. Part of this work included collecting student writing samples—including work at each proficiency level. Students were also able to self-assess their own writing using successful teacher-presented models along with writing rubrics that established in student-friendly language the learning targets (process standards) they had practiced.

We also encouraged her team and other teachers in the department to start using writing vocabulary that aligned with the CCSS ELA and C3 Framework for Social Studies. Over time, we saw how her students consistently and more skillfully began to write effective historical arguments.

This chapter highlights the ways that collaborative teams can make meaningful revisions to instruction to teach students to think like a social studies expert, whether writing general text or for specific disciplinary purposes. First, we relate how teachers we worked with were able to enhance their writing instruction by collaborating with teachers in other disciplines to create a common schoolwide writing vocabulary aligned with language in the CCSS ELA and the C3 Framework for Social Studies. Next, we review the CCSS College and Career Readiness Anchor Standards for Writing and note specific strategies that teachers can apply to support instruction based on these standards. We then describe in detail specific writing strategies collaborative social studies teams can implement to enhance writing instruction in their classrooms. Finally, we offer considerations for addressing students who continue to struggle or for those already showing mastery.

Collaboration Around Consistent Language

Since the social studies field demands high literacy levels, many social studies teachers require students to write regularly. Essays, DBQs, and other forms of writing are common in the social studies classrooms at our school. Although many teachers do provide some lessons on writing, social studies teachers often rely on students having learned writing skills in their ELA classes (as the scenario that leads off this chapter illustrates). However, as students progress as writers through middle school and high school, the differences between English writing and social studies writing increase. In addition, when our literacy team realized that our social studies teachers were using different terminology with their writing requests and instruction than other disciplines within the school, we concluded it was necessary to move toward a common vocabulary.

How did we come to this conclusion and begin the work of adopting common language across our school? Well, at the beginning of the 2017–2018 school year, we sought interested volunteers from all content areas to share their thoughts about writing. We even offered to buy them lunch because food is *always* the best way to attract collaborators and idea sharing! The response was robust, as we found committed teachers from almost all content areas willing to share. The first meeting consisted of teachers from different content areas—social studies, ELA, mathematics, science, and so on—sharing their writing expectations, structures, and samples. Subsequent meetings involved even more teachers from different disciplines examining the different types of writing and finding commonalities.

From those conversations, it became clear that, as a school, we should be consistently using the same terms, such as *claim* (instead of *thesis* or *hypothesis*), *subclaim* (instead of *topic sentence*), *evidence* (instead of *quotes, examples,* or *data*), *reasoning* (instead of *justification* or *elaboration*), and *conclusion*. In this way, as students move from class to class and discipline to discipline, they hear the same language from all their teachers. Yes, there are unique qualities to writing in a mathematics class as opposed to a social studies class, and students need to learn those differences; however, students are able to make stronger connections with what effective writing looks like in different disciplines when they see how a common term (such as *evidence*) is used in different academic contexts. Having teachers discuss these connections between terms and model them in each of their respective disciplines clarifies for students how those terms apply within each discipline.

Happily, when we posed these ideas to teachers on our social studies team, they agreed to stay true to these elements, whenever possible, as they created varied writing activities. Toward the end of the 2017 fall semester, we met again as a group to reflect and discuss results with anyone who had tried to use the common writing language we developed in the spring, including our ELA teachers who had diversified their writing activities while staying consistent with the common language. This conversation was overwhelmingly encouraging. Mathematics, social studies, science, and ELA teachers all reported increased success with student writing once they started using the common language. For example, when social studies students understood that writing a claim was similar to writing a three-point thesis—a focus for their work—they had a better idea about developing clarity and focus in their writing. Similar connections between data and evidence, and reasoning and justification helped students who may have more affinity for other content areas, like science, better understand various language roles in social studies.

Admittedly, content-area teachers did not just come up with the terms *claim*, *evidence*, and *reasoning* on their own. In fact, those terms are key elements of argumentation, and they are embedded in the CCSS ELA (NGA & CCSSO, 2010). One of the first times educators in our school encountered the common argumentation language was in meeting with educator and author Katie McKnight in September 2014. McKnight brought to our attention the work of former teacher Eileen Murphy and her colleagues at ThinkCERCA (n.d.a), a literacy courseware company. (CERCA stands for *claim, evidence, reasoning, counterargument,* and *audience.*) After that 2014 meeting, different teacher and curricular teams began using these terms. Over the next few years, the ideas organically filtered through different divisions and collaborative teams as content-area teachers loosely adopted the terminology.

Some teams made the decision to change some of the language they had been using, such as changing *thesis* to *claim*, and many teams that committed to using the language saw some immediate benefits. First, they realized that the language aligned with the wording used in standards such as the C3 Framework for Social Studies or the CCSS ELA. Second, they shared anecdotal evidence that their students were writing more effectively because of the common language they heard throughout the school day. According to ThinkCERCA (n.d.b) research, "teaching students how to make Claims, support their claims with Evidence, explain their Reasoning, address Counterarguments, and use Audience-appropriate language is the most effective way to improve achievement on assessments and prepare students for post-secondary life." With research and logic behind us, we were happy to facilitate the proliferation of this common language. What started as a loose adoption transformed over time into an organized and collectively determined schoolwide writing vocabulary. The writing vocabulary is still not a mandate from leadership, but there are only a small handful of teams in the building that use unique language; it's a work in progress the PLC continues to hone.

This is all to highlight that, even for social studies teams that collaborate within a PLC culture, there may be gaps around writing instruction. For example, teams may have been working with writing instruction for some time but lack a common instructional approach. Some teachers may emphasize different elements of writing instruction over others. Therefore, an important first step for a collaborative team of social studies teachers is to reach consensus around the first critical question for every PLC: What do we want students to learn? (DuFour et al., 2016). Teams will need to come to consensus on what every student should be able to do as a writer within the core curriculum. From there, teams can plan units with writing instruction in mind. It's important to start small, but that should expand as collaborative teams across the PLC work together to create a common literacy taxonomy that works for them and their schools and help students make literacy connections across curricula.

thinking
BREAK

How can you implement a common language for literacy skills, such as the ThinkCERCA (n.d.b) model or other argumentation tools, in your classroom?

C3 Framework for Social Studies and ELA CCSS Connections

To help develop sound writing strategies for the social studies classroom, it's essential to understand the directive in the CCSS ELA that clearly asks teachers to use the elements within the CERCA (n.d.b) model of argumentation. However, we must keep in mind that social studies experts have an expansive number of modes and reasons for writing that range from producing ideas to analyzing informational texts. The CCSS College and Career Readiness Anchor Standards for Writing are as follows.

- Write arguments to support claims in an analysis of substantive topics or texts using valid reasoning and relevant and sufficient evidence. (W.CCR.1)

- Write informative/explanatory texts to examine and convey complex ideas and information clearly and accurately through the effective selection, organization, and analysis of content. (W.CCR.2)

- Write narratives to develop real or imagined experiences or events using effective technique, well-chosen details and well-structured event sequences. (W.CCR.3)

- Produce clear and coherent writing in which the development, organization, and style are appropriate to task, purpose, and audience. (W.CCR.4)

- Develop and strengthen writing as needed by planning, revising, editing, rewriting, or trying a new approach. (W.CCR.5)

- Use technology, including the Internet, to produce and publish writing and to interact and collaborate with others. (W.CCR.6)

- Conduct short as well as more sustained research projects based on focused questions, demonstrating understanding of the subject under investigation. (W.CCR.7)

- Gather relevant information from multiple print and digital sources, assess the credibility and accuracy of each source, and integrate the information while avoiding plagiarism. (W.CCR.8)

- Draw evidence from literary or informational texts to support analysis, reflection, and research. (W.CCR.9)

- Write routinely over extended time frames (time for research, reflection, and revision) and shorter time frames (a single sitting or a day or two) for a range of tasks, purposes, and audiences. (W.CCR.10; NGA & CCSSO, 2010)

As discussed in the previous chapters, and given the direct connection between the four dimensions in the C3 Framework for Social Studies and CCSS ELA skills, we provide a pair of tables to show connections between this chapter's strategies and these standards. Table 6.1 highlights connections between this chapter's writing strategies and the CCSS ELA skills, while table 6.2 shows the same connections through the lens of the C3 Framework for Social Studies.

Table 6.1: CCSS ELA and Writing Strategy Connections

Writing Strategy	W.CCR.1	W.CCR.2	W.CCR.3	W.CCR.4	W.CCR.5	W.CCR.6	W.CCR.7	W.CCR.8	W.CCR.9	W.CCR.10
Two-Quote Paragraph Template	✓			✓	✓		✓	✓	✓	✓
Evaluate and Support Claims With Evidence	✓				✓				✓	✓
Color-Coded Paragraphs	✓	✓	✓	✓	✓		✓		✓	✓
Rubrics for Students to Evaluate Writing	✓	✓	✓	✓	✓				✓	✓
Historical Sentence Starters	✓	✓		✓	✓		✓		✓	✓

Source for standards: NGA & CCSSO, 2010.

Table 6.2: C3 Framework for Social Studies and Writing Strategy Connections

Writing Strategy	Dimension 1: Developing questions and planning inquiries (R1)	Dimension 2: Applying disciplinary concepts and tools (R1–10)	Dimension 3: Evaluating sources and using evidence (R1–10)	Dimension 4: Communicating conclusions and taking informed action (R1)
Two-Quote Paragraph Template		✓	✓	✓

Evaluate and Support Claims With Evidence		✓	✓	✓
Color-Coded Paragraphs	✓	✓	✓	✓
Rubrics for Students to Evaluate Writing		✓	✓	✓
Historical Sentence Starters	✓		✓	

Source for standards: National Council for the Social Studies, 2017.

If ThinkCERCA (n.d.b) terms do not make sense for your team, what are some common literacy terms you can agree upon to help students connect their literacy skills across the curriculum? How can you implement the ThinkCERCA model or other argumentation tools into your classroom?

thinking
BREAK

Strategies for Supporting Students in Writing

There are many different ways to get students to produce thoughtful writing. Along with utilizing a common writing vocabulary, using *mentor texts* (published writings that students can emulate; Fountas & Pinnell, 2012), creating models based on students' efforts, and cowriting can all lead to wonderful results over time. More often than not, strategies come down to modeling and practice. When possible, work with other teachers to share samples, practice ideas, and give feedback. When you collaborate with different teams within a school community to share the ways you all teach and assess writing, you can make powerful connections around the ways that all of your students engage in different writing tasks in different disciplines. The following strategies and differentiation options for students learning English, students who qualify for special education, and those who show higher-level proficiency, created in conjunction with our social studies teachers, will help you and your team build students' writing competency.

Two-Quote Paragraph Template

Sometimes, structured, formulaic writing is the best way to teach students how to start writing (and thinking) with clarity. This tried-and-true method is an effective tool to lay down strong writing foundations at the beginning of a course. Our social studies teachers have found that using this strategy helps students make connections to writing instruction in basic question responses, essay responses, and a variety of other writing tasks. Called a *two-quote paragraph* because it asks students to include two quotes to support their argument, this strategy offers students a template for constructing effective argumentation. The goal is to eventually break away from the template as students internalize the writing structure; nonetheless, providing students with a sturdy structure for writing helps them ensure they are using strong logic and reasoning in an organized manner.

How to Use

This strategy is most effective when teaching the basics of written argumentation. We recommend using this early in the school year when students are learning how to construct answers to inquiry questions while reading texts. Students will use a template, like the example in figure 6.1, to ensure they are incorporating all three required argumentation elements: (1) claim, (2) evidence, and (3) reasoning. (See page 190 for the reproducible "Two-Quote Paragraph Template.") In this example, a teacher asks ninth- and tenth-grade world history students to write a paragraph identifying the accomplishments of Alexander the Great as an empire builder. The teacher provided them with a variety of primary sources to support their response.

Sentence Number	Type of Sentence
1	**Topic sentence:** State your claim (purpose of the text). Alexander the Great was a successful empire builder because he was able to spread Greek law, and culture across a wide area in Africa, the Middle East, and Asia.
2–3	**Lead-in:** Establish a context for the upcoming evidence or explain a little about your claim. You might define vocabulary or add context to further your idea. As Alexander spread his empire, he encouraged other Greeks to marry people from other cultures. He also established a common currency throughout all his lands. His main goal was to unite everyone.

4	**Quotation or evidence:** Make sure you choose one of the best quotes or pieces of evidence from the text that will support your claim. The Greek scholar Plutarch wrote that Alexander "conducted himself as he did out of a desire to subject all the races in the world to one rule and one form of government, making all mankind a single people."
5–6	**Explanation or reasoning:** Now explain why your evidence is important. How and why does this evidence support your claim? Explain the reasoning behind why your evidence proves your claim. It was important to Alexander to make sure that all of the people under his rule were treated with the same laws and government. His goal was to make "mankind a single people," but that can sometimes be controversial as different people and countries have different values.
7	**Quotation or evidence:** Make sure your second quote or piece of evidence supports your claim as well. Make sure you transition carefully into the evidence—don't just drop it in the middle of the paragraph with no sentence setting it up. Some historians believe that he may not have had good intentions, writing, "That he aimed at world domination is undoubted . . . he probably sought no more than to be king of it all, and sought only to govern as he thought best."
8–9	**Explanation or reasoning:** Now explain why your evidence is important. How does it support your claim? Explain the reasoning why your evidence proves your claim. World domination usually comes at a cost to those being taken over. It's unclear whether he really had good intentions when he was growing his empire. He was effective at growing his empire, but it's debated if it was the best thing for those under his power.
10	**Return to the claim (concluding sentence):** This last sentence should reconnect your evidence and ideas back to your primary claim. Yes, Alexander was able to successfully spread his empire across several continents in a short period of time, but some historians debate whether he had good intentions or not.

Figure 6.1: Sample two-quote paragraph template.

As you can see from figure 6.1, the template provides a clear, sentence-by-sentence structure for students to follow, and you can easily adapt it to meet the needs of the students and the writing task. Is it formulaic? Yes, but that's the point. It helps students learn what good writing looks like. Students often ask, "How long does this have to be?" While the template is flexible, it does help to provide guidance on where students need to elaborate on their ideas.

Adaptations

When thinking about how to adapt this model for students who struggle with this writing strategy, the first thing to consider, as mentioned in previous chapters, is modeling the process multiple times. You will want to offer plenty of samples for each aspect of the process in figure 6.1 (page 140). For example, the more students see examples of well-written topic sentences and claims, the more ideas they will have to draw from when they begin to craft their own. Having students work with more proficient partners also helps clarify ideas that may not be as easy to comprehend. In addition, think about providing specific words you expect the students to use in their writing in each of the template boxes.

Although modeling alone will help all students who struggle with this task, you might also give students who continue to struggle the claim statement and then have them work through the remainder of the strategy. This scaffolding makes it easier to focus the rest of the writing. Ultimately, the goal is for the students to complete the paragraph independently, but this may take a few extra steps to help students learn and understand the process.

Highly proficient students who have mastered paragraph organization may no longer require the use of the template. This is the goal for all students, so challenge proficient writers to produce the same quality and level of work without a template. For those who take longer to demonstrate mastery, have them continue to use the template as needed.

Evaluate and Support Claims With Evidence

This strategy works best for laying the foundations for argumentative writing skills in the social studies classroom. It asks students to evaluate claims and then independently identify evidence that supports the most effective claim statement. This is a powerful formative assessment in that students see a range of claim statements and must then find evidence that best supports each claim. By scaffolding instruction to help students first identify what an effective claim looks like, followed by identifying the most appropriate evidence, teachers break down the writing and thinking process into separate stages.

How to Use

Provide students with three or more claim statements that range in effectiveness, ask them to evaluate each claim, and have them select the best one. Figure 6.2 provides an example from a class on U.S. history. (See page 192 for the reproducible "Evaluating Claim Statements and Supporting Them With Evidence.") After students make their selection, do a formative check and discuss why students made specific selections, or allow students to begin to select evidence for whichever claim they selected. If students choose the most effective claim, they should have little difficulty finding appropriate evidence. If students choose a less effective claim, they might struggle to find evidence, which may require them to rethink their original selection. Require students to identify multiple pieces of evidence and provide reasoning for each selection.

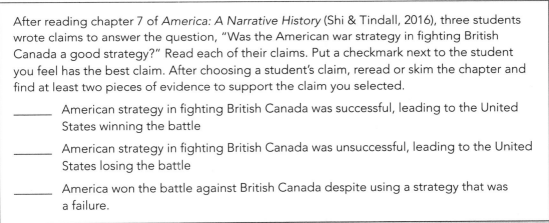

After reading chapter 7 of *America: A Narrative History* (Shi & Tindall, 2016), three students wrote claims to answer the question, "Was the American war strategy in fighting British Canada a good strategy?" Read each of their claims. Put a checkmark next to the student you feel has the best claim. After choosing a student's claim, reread or skim the chapter and find at least two pieces of evidence to support the claim you selected.

_____ American strategy in fighting British Canada was successful, leading to the United States winning the battle

_____ American strategy in fighting British Canada was unsuccessful, leading to the United States losing the battle

_____ America won the battle against British Canada despite using a strategy that was a failure.

Evidence	Page Number	How the Evidence Supports Your Chosen Claim
The invasion was a "disaster" because Americans couldn't field three armies to properly invade in three places.	268	The United States planned to invade British Canada at the same time in three places, which would have required three different army units. Unfortunately, the United States was not able to field three such units, which ultimately doomed the strategy.
"In April 1813, an American force . . . attacked York" forcing British and Canadian troops to surrender.	270	This small victory gave hope to the Americans and their troops. If they could invade and conquer an important place like York (later, Toronto), they would begin to believe they could win the war. This led to other victories along the border, ending up in a stalemate that reduced the threat in the North, which helped the United States eventually win.

Figure 6.2: Evaluating claim statements and supporting them with evidence.

Adaptations

You can adapt this literacy strategy for students who are struggling to master it by giving them the correct claim and focusing them on the search for evidence. Once students have a claim, you may decide to narrow the page range required for these students to reread and search for evidence. Narrowing down the range will help them focus on thinking critically rather than searching through a longer text. As students progress, you may add in the second claim and then the other to build the challenge. You may also reverse this strategy for students by providing the evidence for them and having them then come up with the claim. This helps students process the information and try to fit it all together without having to come up with the claim evidence. Eventually, students will be able to use this strategy without these adaptations, but scaffolding it, at first, will help them work with the ideas.

For students that are learning English, you might add in some frontloaded vocabulary that is important to this specific task. This ensures students can read the desired chapter or passage well enough to perform the evidence search. In addition, make sure teacher-written claims are clear enough for all students to understand.

For students showing high proficiency with this strategy, consider providing a claim that may seem viable, but is provably false. Have students work to *disprove* the questionable claim in addition to proving the viable claims.

Color-Coded Paragraphs

As we stated in the Two-Quote Paragraph Template section (page 140), when students are learning how to write argumentative paragraphs in social studies, they frequently ask, "How much do I need to write?" Additionally, they need guidance to ensure they have all the necessary components of an effective argumentative paragraph. Color-coding drafts of student writing is an effective visual strategy to help students see how each component of the argument contributes to the larger whole. Students will have a clear visual indicator for every component in the paragraph, and by comparing their colored paragraphs with exemplar or peer writing, they can assess their own writing.

Expectations for the product students produce will vary based on the writing task, so teachers should clearly indicate to students what an appropriate ratio of color codes looks like based on the assignment. As for a balanced end product, students should have fewer sentences for the claim section, and more content to provide evidence and support their reasoning. Again, this may vary depending on the individual assignment.

How to Use

Assign a color for every required component of a writing task (for example, *claims* are red, *evidence* is yellow, *reasoning* is blue, and so on). Students then use a highlighter or a word-processing application to highlight the entire paragraph, applying the designated colors to each paragraph component. Students compare their own highlights with an exemplar or peer's work to ensure that the quantity and ratios of sentences for each of the areas (claims, evidence, reasoning, and so on) of their writing are sufficient. Students also work together to discuss differences of opinion in how they assessed the text. This is not an effective strategy for assessing the *quality* of the writing, but it does ensure that students have the necessary components. It's also a good strategy to use as students begin to break away from the two-quote paragraph template (page 140).

As an example, figure 6.3 shows an excerpt from a student's final paragraph supporting her identification of the argument in the article "'Hi! My Name's Eva': A Teenage Holocaust Victim's Diary Comes to Life on Instagram" (Eglash, 2019), which might be used in a world history course or in a unit on World War II and the Holocaust. The sample is annotated as follows: the claim is shaded in pink (light gray on this page), the evidence is shaded yellow (medium gray), and the reasoning is shaded in blue (dark gray). In this instance, the student is able to see that her writing has all of the necessary elements for this writing task and that it has an adequate amount of writing for the given task.

> The author suggests that using a new media format to raise awareness of the Holocaust in today's world of short attention spans is effective. Eglash concludes that the Instagram story launch has been seen by more than one hundred million people and has been successful by "not only going viral online but also sparking conversation among Israeli youths" (Eglash, 2019). Considering so many people are viewing and talking about a real-life story from the Holocaust, this new way of preserving history is effective and will keep Holocaust stories alive for future generations.

Figure 6.3: Color-coded excerpt from student writing.

Adaptations

This visual-based strategy is especially helpful when used with students who struggle with writing because it gives them a clear and concrete key to understand what they are writing and ensures they are hitting all of the components of their writing in social studies. If students feel overwhelmed by focusing on multiple elements of their writing, teachers can scaffold the instruction by having them highlight one element at a time (focusing on one skill at a time). Alternatively, teachers may provide a completed paragraph with each portion cut into paper strips that students can manipulate to place in the correct order. Students can then label the different parts of the paragraph using the argumentation terms.

For students learning English whose language barriers may cause struggles with incorporating all the different writing elements, teachers may opt to use word banks or sentence starters to facilitate writing. Prior to color-coding their paragraphs, students can reference the sentence starters or task-specific words to support their writing. For example, a student struggling to incorporate textual evidence can choose from a number of leading phrases, such as "The text states . . ." "The author argues . . ." or "One expert declared . . ." to scaffold their efforts on the way to producing grade- or course-level writing.

Similar to previous writing supports, like the two-quote paragraph template, students that have consistently demonstrated proficiency with writing organization should not have to use this strategy. As more students reach mastery with writing organization, they can begin to break away from some of these scaffolded supports.

Rubrics for Students to Evaluate Writing

Many schools, especially those that operate as a PLC, have shifted or are shifting toward standards-based assessment and grading (Townsley & Wear, 2020). According to educator Danielle Iamarino (2014):

> Standards-based grading focuses . . . on larger outcomes; rather than inferring a student's progress based solely on how many points the student has accumulated from the completion of individual assignments—or from attendance, which is sometimes scored alongside academic assignments—standards-based grading concerns itself with the cohesive body of knowledge that the student gains as a result of the course. (p. 1)

Because this approach helps students focus on the final results, using standards-based rubrics for evaluating writing is effective in developing students' working knowledge of writing skills. To help students learn to focus on acquiring the writing skills necessary to meet course learning standards, team members can work collaboratively to provide students with sample writings and standards-based assessment rubrics to coach students to understand, learn, and interact with the skills necessary for success.

While learning and being coached to understand the targeted skills, students have an opportunity to reflect on their own understanding of success or mastery of a given skill. This close work with written samples and the targeted skills on the rubric helps students to be successful in transferring understanding to their own writing habits.

How to Use

Prior to a writing assessment, give students the opportunity to assess teacher team–created writing samples or past student writing samples embedded within a rubric. The rubric should be the same tool that you intend to use to assess the final outcomes of students' work on the eventual writing assignment. The writing samples the team collaboratively selects or creates should include both claims and body paragraphs for a writing assignment or other products related to the specific assignment. Using the standards-based language from the rubric, students then assess the teacher-provided samples by identifying strengths and areas for growth. Students can do this individually or in small groups; however, it is best to incorporate a range of sample paragraphs so students have a clear indication of what effective *and* ineffective writing looks like.

Figure 6.4 (page 148) provides an example of a U.S. history lesson where students evaluated two different body paragraphs related to an essential question: To what extent did territorial expansion unite and divide Americans from 1800 to 1860? In this instance, the teacher used the lesson to debrief a recent writing assessment using past student samples. The lesson allowed for the class to come to a consensus on what makes an effective body paragraph, and in a follow-up lesson, the students used the same criteria to assess their own writing. (See page 193 for the reproducible "Rubric to Evaluate a Writing Sample.")

Using rubrics for standards-based assessment and grading also expedites the grading process for teachers. By providing a clear range of writing criteria for students, teachers do not need to spend as much time writing personalized feedback for every student. The scoring criteria within the rubric inherently provide feedback by highlighting what elements a student's writing either included or lacked. This facilitates a learning environment in which students take greater ownership throughout the writing process, helping them reflect on their own writing strengths and areas for growth.

Subject: U.S. History

Student name: Aliyah Ubben

Directions: Read and reflect on the following essential question. Then, read each sample claim statement and body paragraph and assess the body paragraphs using the Historical Evidence and Supporting Argument portion of the embedded rubrics. Beneath each rubric, explain the score using the language of the rubric.

Essential question: To what extent did territorial expansion unite and divide Americans from 1800 to 1860?

Sample A (claim): Although some believe that territorial expansion united Americans, Westward Expansion toward the Pacific greatly divided America from 1800 to 1860.

Body paragraph: As the number of cotton plantations increased, the Southern economy was boosted. However, the need for slave labor increased as well. This was known as slavery expansion, which divided the Northerners and Southerners. The Northerners believed that expanding slavery was wrong since it was against their morals. Southerners believed that increasing slavery was the right thing because they feared that the Northerners would have enough votes in the U.S. Senate to outlaw slavery. Slavery expansion would have a role in the outbreak of the Civil War. The Louisiana Purchase was another example of how territorial expansion greatly divided America. To pay for his European conquest, Napoleon offered to sell the Louisiana Territory and New Orleans to America. On April 30, 1803, with Congress's approval, President Thomas Jefferson bought the land for $15 million. While the purchase would double the size of the United States, it also created a dilemma for Jefferson. The U.S. Constitution doesn't give the president the power to buy land. The purchase also made many believe that the land was a waste of money and unnecessary. As a result, territorial expansion greatly divided Americans from 1800 to 1860.

Writing Criteria	Scoring Criteria			
	1—Needs Improvement	2—Approaching Mastery	3—Mastery	4—Exceeds Mastery
Historical Evidence (Factual Information)	Inappropriate or no use of evidence	Little use of evidence or information is inaccurate or irrelevant ✓	Clearly states and generally defines evidence to support the claim (argument)	Clearly states and thoroughly defines evidence to support the claim
Supporting Argument (Analysis)	Inadequate or inaccurate analysis of the evidence	Has evidence, but analysis may be limited or simplistic ✓	Includes evidence, and, in most cases, explains how it supports the claim	All supporting evidence specifically and consistently explains how it supports the claim.

Explanation:

The writer is approaching mastery in use of historical evidence because evidence is stated but not fully documented. Some of it was opinion based, such as saying the North didn't like slavery, but they don't back it up with specific evidence. This is also why I scored the analysis as criteria as approaching mastery, because it does not use source material or quotes to support the claim. Specific evidence would help to make it less simplistic. The claim does not specifically write about uniting and dividing America.

Sample B (claim): Territorial expansion opened conflict within the country. Topics of slavery and citizens' rights created a division between the North and South. However, some events, like the Market Revolution and the Seneca Falls Convention, united the country through easier trade and bringing attention to the grievances of women.

Body paragraph: Territorial expansion divided the country because different groups in the country—Northerners, Southerners, and the U.S. government—had conflicting views on various topics, specifically slavery expansion and the Louisiana Purchase. After the invention of the cotton gin, Southerners were able to produce more products, which led to increased demand for slaves. This created a conflict between the South and North because the South saw slave labor as a valuable asset to their way of life, but Northerners believed that slavery was wrong and should be abolished. Additionally, the Louisiana Purchase divided the government from the citizens due to the questionable constitutionality of President Jefferson's actions. The citizens believed that the president had no power under the U.S. Constitution to buy land, but since the Constitution did not specifically state that the President does not have the power to buy land, Jefferson did not see this purchase as unconstitutional. Although there were many actions that caused division in America, some events also unified citizens.

Writing Criteria	Scoring Criteria			
	1—Needs Improvement	2—Approaching Mastery	3—Mastery	4—Exceeds Mastery
Historical Evidence (Factual Information)	Inappropriate or no use of evidence	Makes little use of evidence or information is inaccurate or irrelevant ✓	Clearly states and generally defines evidence to support the claim (argument)	Clearly states and thoroughly defines evidence to support the claim
Supporting Argument (Analysis)	Inadequate or inaccurate analysis of the evidence	Introduces evidence, but analysis may be limited or simplistic	Includes evidence and, in most cases, explains how it supports the claim ✓	Includes evidence that always specifically and consistently explains how it supports the claim

Explanation:

This writer is approaching mastery in the use of historical evidence because they are specifically mentioning historical events, but they don't give as much evidence as they should. For example, they write "Jefferson did not see this purchase as unconstitutional," but it's not backed up by source material. The analysis is mastery level because the writer does a nice job addressing both sides of the prompt by analyzing how it united and divided America in this period. They also give specific examples of historical events.

Figure 6.4: Using a rubric to evaluate a writing sample.

Adaptations

When evaluating writing, it may be difficult for students who struggle with writing proficiency to identify the degree to which writing aligns to specific scoring standards in the rubric. Teachers can help students color-code the different writing samples to help them to understand what is considered historical evidence (facts) or supporting arguments (analysis), which helps them understand how they align to scoring criteria in the rubric. This ultimately helps students measure how well the writing samples align to learning targets. This is very similar to the color-coded paragraphs strategy (page 144), and teachers can use color-coding in this way throughout the year as a commonplace aspect of the work students do in the classroom.

For students learning English, it is essential to ensure their access to the rubric vocabulary. While this academic vocabulary should be something that they are working on in many of their classes, you want to ensure they understand the writing and scoring criteria necessary for success.

To extend learning for students who exhibit proficiency in evaluation, have them improve on the sample writing for areas they identify as needing work to reach mastery. Including this step in the activity gives students practice for revising their own writing.

Historical Sentence Starters

If students are to write like social studies experts, they need the necessary academic vocabulary to articulate their thinking. There are a number of skills that are essential to thinking like a disciplinary expert in the realm of social studies, such as identifying cause-and-effect relationships, interpretation of sources, or evaluating various perspectives and subjective data. Providing sentence starters that teachers categorize into different modes of thinking and analysis helps students use their writing to communicate like a disciplinary insider. Teachers can use the sentence starters to focus on specific skills, like claim writing, topic-sentence writing, and so on, or they can use them to help think more broadly about a given topic, such as a starting point for a longer research paper or report.

How to Use

Provide a list of sentence starters broken down by category. These categories will apply to the types of sentences required in any given assignment, be it a document-based question, essay, or another type of writing. It may be best to scaffold a few of these sentence starters over time so as not to overwhelm students. Make sure to meet your students where they are at and consider what they can manage. You might start them off with three to five starting options. Then, as you move along with different prompts and assignments, keep offering a couple of new options. Students may continue to use the first set of starters, but you can also use the additional options to push for variety and sophistication as the students get better at taking the prompts and running in new and original directions with them. This can also work well as a living document that students add to throughout the year. Display the starters on a poster in the classroom or give students access to an electronic document with reminders to consult it as needed. What matters is that teachers and students consistently return to the sentence starters as students engage in the writing process. Over time, students should start to internalize the sentence stems as they begin to master writing skills for social studies. Over time, they will be able to work just as effectively (or even more so) without them. Figure 6.5 provides an example of sentence starters you might provide, broken down by category. (See page 195 for the reproducible "Historical Sentence Starters.")

Cause and Effect

There were economic, political, and social factors that contributed to . . .

The most likely reason that . . .

Several factors contributed to _____ , such as . . .

_____ led to _____ . . .

_____ had a significant impact on _____ because . . .

_____ was a result of _____ . . .

Interpretation or Perspective

These actions changed the course of _____ because . . .

This was a significant event because . . .

One interpretation of this is. . . .

Others argue that . . .

The evidence suggests . . .

One way to interpret this event is . . .

Examining the event through the perspective of . . .

Figure 6.5: Historical sentence starters.

continued ⟶

Sourcing

The author may be motivated by . . .

Even though the author says _____, he or she may mean . . .

Based on the era of this source . . .

At the time of this source, _____ was occurring, so . . .

The author may be biased because . . .

Adaptations

This is an excellent strategy for use with all students who struggle getting started with their writing assignments, but it's an especially useful scaffolding strategy for students in special education and students learning English. Have students who need additional support practice using the same sentence starter multiple times. As students' comfort grows with a specific starter, it will benefit their ability to expand their learning using subsequent starters, eventually not needing the support at all.

As with most strategies, discipline-specific vocabulary knowledge is essential for students learning English as they approach writing assignments. If students do not understand the vocabulary in the sentence starters, they will struggle with the activity. Make sure that difficult vocabulary is frontloaded either with a vocabulary strategy lesson, such as Frontloading Vocabulary With the Frayer Model (page 63) or with a glossary. In addition, it may be useful to provide a word bank students can use for filling in the blanks in the sentence starters.

Proficient students are unlikely to need the full modeled starters and are therefore better served attempting to write sophisticated content without these scaffolds. Instead of providing proficient students with full starters, give them a word bank of key words and phrases they can choose to combine as part of their writing assignment.

Considerations When Students Struggle

There are many complexities inherent within the social studies discipline that challenge students' writing skills, and students will undoubtedly need help and support mastering these skills, from getting started with a piece to disciplinary vocabulary and grammar to content and organization. We know how incredibly frustrating it is for teachers (as well as students and parents) when students have ideas for what they want to say but struggle to articulate them through writing.

With that in mind, here are some additional questions to consider when you encounter students who are having a hard time writing.

▸ **How might you break down tasks to allow students to focus on one idea at a time?** Consider how your team can work to develop a scope and sequence of writing skills horizontally (at a specific grade level) and vertically (across a range of grade levels). It can be beneficial to focus instruction around one or two skills at a time. For example, focus on having all eighth-grade students master writing claims. In ninth grade, set a goal that all social studies students should review claim writing and build in finding appropriate evidence. Also, consider the ways in which your team provides feedback to students. Don't feel the need to mark everything! Use rubrics and focus feedback on what matters most: demonstrating proficiency with targeted writing standards. Students can only process so much feedback for every writing experience.

▸ **How might you help students write first and worry about editing later?** Consider some of the ways students engage in the writing process. Do students feel like they have opportunities to take risks and make mistakes? In our experience, students experience writing paralysis (an inability to write at all) when they feel like they need to be perfect. Since grammar and writing conventions are likely not your top priority as a social studies teacher, focus on what matters most—content—and then give students opportunities to edit their work when they have confidence in their written content.

▸ **How might you find time to confer with students one on one to help them before they write, while they write, or after they write?** This may vary depending on the assignment. For example, you might initially support a student before or during his or her writing. As the student's writing ability grows, you might move that support to a more reflective role after they complete their work. Working with the students in this way provides a model for future work and steers students toward completing writing assignments independently. It may also be helpful to limit conference time to focusing on one skill at a time so you can quickly meet with as many students as possible. For example, spending two minutes per student, specifically looking at each student's claim, allows you to meet with several students during a single class period while still providing individual support.

▸ **How might you pair students together to build on each other's writing strengths?** When teachers create a community of writers within the classroom, students can learn from each other. Peer feedback is a common practice in ELA classrooms, and it can be just as effective in social studies classrooms. However, unless students learn how to give effective feedback, it may not be a meaningful experience. Therefore, it's important to provide students with examples of effective and ineffective feedback and give students repeated opportunities to support each other's writing.

Because writing is a complex task, teams will benefit by identifying one or two concrete skills to focus on for collaboration and instruction. As teams (and students) begin to master and build on those skills, they can refocus attention on other skills, coming to consensus on instructional strategies like the ones outlined in this chapter. As teams experiment with different instructional approaches to complex writing tasks, each team will continuously increase its capacity to support students as social studies writers.

thinking
BREAK

Review the strategies in this chapter. How might you help students who struggle to make progress?

Considerations When Students Are Proficient

Students who have demonstrated proficiency in writing still have room for growth. Here are some suggestions to support students who are ready to take on more complex writing tasks.

▸ Encourage proficient writers to increase synthesis skills by asking them to embed multiple sources or even a mix of current sources and primary documents into their writing. This allows students to make connections within and beyond social studies units.

▸ Experiment with different sentence variety, perhaps encouraging students to emulate published historians.

▸ Increase the use of more technical social studies and historiographical vocabulary, as relevant to the unit of study.

Wrapping Up

Reading and writing go hand in hand. More often than not, growth in one area can lead to growth in the other. Therefore, it is important to provide consistent opportunities for students to make connections between their reading and writing experiences. When you confer, model, and share student samples with the students in your classes, try to provide feedback that is both encouraging and constructive. While your teacher team plans writing instruction, consider how to prioritize writing skills that meet the needs of your students and content area. As the school year progresses and the team collects more evidence of writing skills, teams should be willing to adapt instruction so that it continues to support the students using best practices decided by the team.

Collaborative Considerations *for* Teams

- Is your social studies team equipped with a writing vocabulary that aligns with the language from your school's standards, such as CCSS ELA and the C3 Framework for Social Studies?

- Are there opportunities for your social studies team to collaborate with teams from other disciplines? Who can facilitate those conversations?

- Review the writing strategies in this chapter. Consider the following questions.

  - What writing-instruction needs do your students have?

  - What strategies from this chapter might best address those needs?

  - Can the team develop mentor texts and models to provide exemplars of effective writing?

CHAPTER 7

Assessment

Many of our team's literacy coaching sessions involve detailed, lengthy discussion on what exactly we are trying to assess. Are we assessing content knowledge? Or, are we assessing literacy skills? Ultimately, the answer is both. If teams want students to grow and succeed, teachers need to hold each other accountable for ensuring all students receive feedback, which includes assessing their work and committing to not only careful and intentional content and skill instruction, but also strategic assessment that tackles these learning goals.

During our sessions, we collaboratively set out to find ways to make most literacy assessments quick and seamless but still informative. Our goal was to ensure that disciplinary content is not lost while also ensuring assessment of proficiency with power (essential) standards and skill development related to literacy. We also wanted to ensure any growth of these skills transfers beyond the assessment.

Together, our team found that through quick formative checks and feedback, we could assess students' literacy work and produce disciplinary growth. Sometimes the assessments came in the form of warm-up activities, like an entrance slip on a previously investigated social studies concept (see page 125), or as a different postreading application strategy, such as the 3–2–1 strategy (page 127). Sometimes the assessments were simple visual scans of whether students ultimately agree or disagree with a statement on the anticipation guide activity from chapter 3 (page 56). And sometimes, teachers had to embed the assessments in more developed thinking activities, like analysis papers or critical reader responses that may take more time to grade but induce students to more clearly articulate content targets while demonstrating literacy-skill growth. The challenge (and the fun) of working as a collaborative team ensures that, from start to finish, formative assessments are useful for both literacy and assessment purposes. Likewise, teachers feel comfortable implementing such activities in their classes, not only because they created them together, but because they know the team will collaborate to assess the resulting data and use the data to further instruction.

There are seven considerations to keep in mind when approaching assessment: (1) understanding the role of literacy-based assessment, (2) matching text complexity and reader capacity, (3) monitoring student perceptions, (4) collaborating to create assessments, (5) using rubrics as assessment tools, (6) providing timely and effective feedback, and (7) analyzing data. In this chapter, we break with the format we established in previous chapters to focus specifically on each of these seven considerations, as teams work to design a variety of assessments that measure student learning in the social studies classroom.

Understanding the Role of Literacy-Based Assessment in the Social Studies Classroom

Assessment is a natural component to the classroom and always has been the primary source of identifying a student's understanding of content. Assessments drive many elements of learning within schools and beyond, including determining final grades and influencing college admissions, which often lead schools, media, and society to focus heavily on this educational tool. The importance that assessments bear means that, as an educator, your school leadership and community stakeholders expect your team to be up to date on the most recent assessment trends and subscribe to some sort of methodology. This task can be daunting; whether it be assessment *for* learning (Stiggins, 2005) or standards-based grading (Townsley & Wear, 2020), there is always a new idea out there.

While there are many assessment philosophies, the consistent element among most of them is using assessment data to inform instruction (Bailey & Jakicic, 2012). If teams are not using assessment to better identify next steps for instruction, they are misusing the data those assessments produce. After reviewing students' assessments, teachers may respond with a follow-up lesson by reteaching a concept the next day or by conducting a long-term response involving curriculum revision. Next steps depend on the purpose and format of the assessment. Typically, most assessments fit into one of three categories.

1. **Formative:** These are in-class assessments that teachers specifically craft to inform themselves and their students of each student's progress toward mastery. These tests may carry score weight but are not the last opportunity for students to demonstrate their knowledge of a target, standard,

or task. Note that depending on your school's testing philosophy, formal progress monitoring (between benchmark periods) may also fall within this category.

2. **Summative:** These assessments are less frequent and serve as an endcap to teaching and practicing a concept. You may administer summative assessments as part of an end-of-unit assessment or even as part of a larger context, like final exams for the semester or school year. That said, as with all team-issued common assessments, we encourage you to use the results, including summative assessments, formatively to drive continuous improvement in teaching and learning.

3. **Benchmark:** Teams offer benchmark assessments at specific intervals throughout the school year to capture student academic growth from a holistic perspective. Many teams use benchmark assessments to gauge students' knowledge at the start of a semester, school year, or high school career, and then again at the end. Some include a benchmark assessment at the midpoint of these as well. These data are typically for tracking students' learning progress and not for determining grades for a particular course. These assessments are often group-administered and nationally normed, meaning that benchmark assessments deliver a percentile comparing each student to others in the same grade or of the same age within a statewide or national pool of students representing all student populations.

As your team approaches the work of literacy assessment within the social studies context, keep the four critical questions of a PLC in mind. These will guide your team discussions and point to your next steps. You may employ these questions holistically to an assessment or to each assessment component. As you do so, trends will emerge, once again pointing your team toward its next step, which may range from reteaching a specific lesson in the classroom to reteaching with a small group of students to engaging one on one with a specific struggling learner. Additionally, your team's findings may point it toward the next level of inquiry. This may occur in the form of conducting a deeper-dive assessment, aiming your instruction, implementing additional student practice, or conducting further assessment around a specific and more narrow skill that a previous assessment identified as a common relative weakness for a cohort of students.

Matching Text Complexity and Reader Capacity

When crafting literacy-based assessments for the classroom, it is vital that your team design with text complexity and reader capacity in mind. First, team members must know the readers in their classroom and their skill sets. Administering an assessment that provides a Lexile measure for each student is a practical way to get this information. Typically, teams design or choose these assessments to administer to a large group of students at one time, often electronically, which enables immediate results. Given RTI (Buffum, Mattos, & Malone, 2018) and MTSS (National Center on Intensive Intervention, n.d.) have become common practice across the United States, many schools already conduct this type of benchmarking assessment; however, time and again, we have noted that the results of such benchmarking often go unused. Reach out to your school's reading specialist to learn more about what type of literacy-based benchmarking assessment might already be in place. If you don't have a building-based reading specialist, check with your principal, who should be able to point you in the right direction.

To ensure schools use these results effectively, leadership teams should consider the following questions.

▸ Does the school conduct benchmark literacy assessments?

▸ When does benchmark testing occur?

▸ Which student populations participate?

▸ Where are the data housed?

▸ What happens with the data?

▸ How can leadership teams share these data with core departments?

Note that anyone can start this conversation. Bringing these questions to your leadership is simply one way to initiate and request the support that your team needs and also prompt your leadership to partake in reflection.

Because all content-area teachers are teachers of literacy, it is imperative that your team accesses students' literacy data. The data will help you scaffold instruction and select materials and texts. Additionally, having comprehensive data will help team members gain a better understanding of why students are striving or struggling. The goal is to understand each student's *independent, instructional,* and *frustration* reading-comprehension levels, as follows.

▶ **Independent:** This is the level of text that a student can navigate independently with fluency and solid comprehension. The student does not need any supports or scaffolds when engaging with this text. A student still learns and gains new insights with these texts, but more likely through learning new background knowledge and making inferences, not necessarily by expanding his or her working vocabulary. Matching a student with an independent-level text is a great option for independent-reading tasks.

▶ **Instructional:** This level of text is appropriate for reading materials teams use in the classroom to challenge students. This level fosters literacy-skill growth, given that teachers provide support to students during the comprehension reading task. A student's instructional reading range falls between the independent and frustration levels and can often span multiple grades of text complexity; for example, a student's instructional reading range may span grades 8–10.

▶ **Frustration:** A text at a student's level of frustration is one where the student has large gaps in understanding due to not having sufficient background knowledge and vocabulary skills to decipher meaning from the text. Employing such a text in the classroom is best reserved for one-on-one or small-group teacher-guided reading with heavy vocabulary and background knowledge frontloading (see the strategies in chapter 3, page 45).

The purpose of knowing each student's level is to use it as your team selects texts, builds a curriculum, and crafts assessments. Clearly, there should be a match in text complexity between lessons and assessments. If your team knows that it has students with a sixth-grade instructional reading level, then using a textbook with a tenth-grade reading level would be at their frustration level of reading. We see this kind of mismatch between student reading levels and textbook complexity quite often when teachers are unaware of the drastic mismatch that often happens with readers and unsupported text. The result is a missed opportunity for skill and vocabulary development, leading to frustrated students and closed books.

As teachers, we have honed and developed strong literacy skills, and as we are teaching high school courses, we often assume that our students are on the same track. However, the reality is that many students cannot work with grade-level materials yet. This reality is sometimes masked as confusion over the content when the roots of the issue are weakened literacy skills, poor vocabulary, and limited background knowledge. By sticking only with grade-level materials, your team misses opportunities to build the vocabulary and background knowledge that your students need to move *toward* grade-level reading comprehension.

Although many of your textbooks might be well above the instructional reading levels of your students, the solution cannot be to abandon a mandated textbook and summarize it for your students. In a PLC, all students need grade- and course-level instruction (DuFour et al., 2016; Mattos et al., 2018). We have seen this trend in the past as teachers moved away from books and toward slide-based presentations (such as PowerPoint) as a means of text summary. Instead, we encourage you to frontload content with interesting, relevant texts that help scaffold background knowledge and student vocabulary. In doing so, your students are more prepared to tackle that challenging grade- and course-level text in class, with teachers there to support them—not at home, left with their own frustrations.

Monitoring Student Perceptions

Assessing students' perceptions of their literacy skills and practices provides insight into how our students approach texts and navigate comprehension pitfalls. In 2018, our school's literacy team decided that, as part of its school improvement work, it wanted to survey students to see how they engage in text for academic and nonacademic purposes. We administered a survey in student-friendly language that asked about specific strategies and the scenarios in which they would apply these skills. On a schoolwide level, the data were certainly interesting, but on a departmental level, the results had even more meaning. For example, annotation is something that particular departments viewed as essential to mastering their content, but few students reported that they did this as a during-reading activity. Additionally, most students reported that, when they feel confused while reading a text, they reread. This sounds great, but it made us wonder—do students know how to reread? Or, are they simply just taking the same faulty approach that led to their confusion in the first place? These two examples have led us to the important next steps: our literacy team needed to work with content-area teams to model fix-up strategies for students and scaffold self-remediation skills.

Collaborating to Create Assessments

The work of designing assessments, including literacy-based assessments, is best done as part of a collaborative team. Doing so, in some capacity, helps teams to write common assessments that accurately measure student growth with content (learning goals) and process mastery while also producing useful data for team discussion. It is important to assess student growth in both of these domains. It's obviously important for students to achieve mastery of learning goals, but it's process mastery that helps students achieve deeper understandings and move toward proficiency with disciplinary literacy.

The texts teams select to use for assessments should speak directly to your team's curriculum standards and echo its formative text tasks. In crafting reading skill inventories, your team must ensure that the text topic mirrors or extends course content. So if you are studying a unit on the civil rights movement and you are using primary sources from that era as your core mentor texts, you should aim to use the same type of text for your team-crafted reading skill inventories. The text and tasks of the reading skill assessment, both formative and summative, should mirror those that you will work on throughout your curriculum. Additionally, the questions and analysis tasks should directly speak to the literacy outcomes that your team's discipline requires. Make sure that the texts are also within an appropriate instructional range for the students your team is testing.

Typically, teams should collaboratively design two or three assessments that target the same skills and either give them as a preassessment and postassessment to monitor prior knowledge and then student mastery of learning goals or as a preassessment, mid-unit assessment, and postassessment to additionally check in on student progress. It is important teachers give the preassessment prior to introducing any skills to establish baseline data and help determine the level of support students need. Teachers should administer the next assessment after there has been time for thorough instruction, guided practice paired with checks for understanding, and independent application. These assessment results will provide ample discussion opportunities for your team to shape how each team member teachers power standards (see chapter 1, page 17).

Know that an assessment does not need to take on one specific format. In fact, using a variety of assessment methods will convey better the extent to which students can transfer and transport content-area literacy skills from one context to the next. It is also essential that your assessments are balanced, meaning they should not all be formative or all summative. Additionally, depending on the structure of your course, not all assessments need to be common. Teams may choose to assess common core content targets, but individual teachers can determine a cluster of different skills to assess depending on the proficiency students show in each classroom. This will foster a truly differentiated learning environment that aligns with the loose-tight nature of a PLC (DuFour et al., 2016). Further, although formative assessments should be your primary tool, this does not necessarily mean that all formative assessments should receive a grade. What matters most is that you are continually providing students with opportunities to demonstrate their learning in incremental phases, leading toward summative assessment tasks. In turn, your team collects data it can use to drive instruction, including intervention and extension.

Strong, literacy-based assessments prompt students to employ the same literacy skills they have been practicing in prereading, during-reading, and postreading activities but with a new text that is of a familiar format. For example, if you have been teaching students to identify the main idea of a primary source in a scaffolded fashion, then you should assess this same skill in a similar fashion. If you are teaching struggling readers, you may even prompt students to employ the same scaffolds you taught and modeled in class; over time, you can and should eliminate these scaffolds as students gain independence and autonomy as disciplinary readers. In such a scenario, the strategies we outline in this book not only serve as your strategy scaffold for teaching a specific literacy skill but also as your assessment of the student's ability to apply that learned skill to new contexts.

Using Rubrics as Assessment Tools

Rubrics have long been a useful tool for teachers to provide students with specific, targeted feedback on their work. When given to students for self-reflection, rubrics can serve as a powerful self-assessment tool. There are many different rubric variations based on the specific target or task the rubric focuses students on. We've seen rubrics commonly used to detail a *minimum* of twenty descriptor boxes, each indicating a level of correctness. Although we sometimes still find these useful for a final, summative assessment, our team has moved toward using single-target rubrics as part of the formative and reflective learning processes.

The sample rubric in figure 7.1 focuses on a single target. (See page 193 for the reproducible "Rubric to Evaluate a Writing Sample.") In this case, the rubric details one target on a proficiency scale (level 3, indicating proficiency with the learning target), with specific success criteria, and provides a space for student reflection and teacher feedback. By providing this rubric to students when presenting the assignment or prior to the assessment, you allow students to know exactly the skills they must engage with to complete a task successfully. Additionally, you can have students complete the reflection column before submitting their work to add an extra layer of self-editing; alternatively, prompt reflection by having students complete it after seeing scores and teacher feedback. Following this with another opportunity to demonstrate skill development on a new practice task or assessment gives students a specific and concrete aspect of the task on which to focus their efforts.

Name:			
Unit: World religions			
Targets: Freshman social studies—literacy rubric for world religions textbook, chapter 3			
4	**(3)**	**2**	**1**
I always read and understand a variety of texts by effectively annotating or taking notes, asking questions, and applying new understanding to other contexts. I can apply and connect knowledge between thematic units. I am able to help my peers by teaching and evaluating their reading strategies.	I can consistently read and understand a variety of texts by effectively annotating or taking notes, asking questions, and making connections that demonstrate this understanding. I recognize when I do not understand, and I ask specific questions to seek support.	I read and understand a variety of texts by taking notes, asking questions, and identifying new understandings. Occasionally, I recognize when I do not understand, and I may ask questions or seek support. I make some connections between concepts within a reading but struggle sometimes to connect ideas outside of a text.	I read for completion of a reading assignment. I cannot identify relationships of concepts within the reading. I do not observe all aspects of the text and struggle to have a complete visual in my mind when reading. I do not realize when my understanding stops, and I do not ask questions to seek support.

Success Criteria Teacher circles one: M = Mastery NI = Needs Improvement	How Well Am I Doing? Student identifies one strength (+) and one area for improvement (–).	Teacher Feedback Teacher circles or highlights successfully met criteria.
Annotation and Notetaking Skills M (NI)	+ I wrote a lot of annotations in the margin. − Most of my annotations are just connections to my personal life.	Notes focus on breaking down reading and showing your understanding; notes connect to the "guts" or main idea of the text Other:

Figure 7.1: Rubric to provide timely and effective feedback.

continued ⟶

Engagement With the Reading (M) NI	+ A lot of what I read reminds me of my trips visiting family outside of the United States.	Demonstrates an understanding of the relationships between ideas by obtaining important information, asking questions, and defining issues or problems Other: *Now try to bring what you read and your personal experiences together. What new understandings do you have?*
Application of Reading (M) NI	+ When I'm lost in the reading and my mind wanders, I know to go back to where the movie in my head was "playing." − I rush and speed read sometimes. I don't annotate those parts.	*Applies ideas in a variety of ways including drawing, graphing, communicating, and so on* Other:

In this example, a student has provided his own self-reflection on his reading of a textbook chapter on world religions; the teacher has reviewed it and provided written feedback about his successes and next steps for improvement. As the teacher, it's critical to observe both the student's approach to reading and the products (annotations, written response, and so on) he or she generates so that you can determine an evaluation in the first column and fill out the third column when providing feedback. Notice how, in the third column, the teacher has circled specific language to indicate what the student is doing well and provided specific notes in the Other field to indicate next steps. As an alternative, if the teacher fills out the first and third columns before the student engages in self-assessment, the student can put the teacher's feedback to use and then use the second column to reflect on how he or she will revise or incorporate feedback into subsequent work.

Providing Timely and Effective Feedback

Feedback is essential for student growth. We're sure everyone remembers a time when they were a student and worked really hard on an assignment or assessment, such as a research paper or a particularly critical final exam. As soon as you handed in your work, all you could think about that day was, How did I do? The very next day, you probably walked into class, just hoping that you would have an answer, any answer, to that question so that your nerves could relax.

We are not suggesting that your teams need to provide *immediate* feedback to students on every assignment or assessment. What matters is finding a balance in the feedback process so that team members prompt student reflection and growth while not getting bogged down in the infinite world of feedback. When teachers are unbalanced in their approach, they often find themselves in a situation where they spend more time providing feedback than the student did completing the assignment. This is not a productive feedback system! Our experience finds that most secondary-level students are simply not able to recognize their own errors, and if they can, they likely don't have the tools to correct their work without some form of prompting. This prompting can happen via a number of discourses, both formal and informal; however, it is critical to recognize that the longer students go without feedback, the less likely they are to reflect on that feedback and learn from it.

While some assignments, like a major research paper, certainly require some time for team members to provide feedback, there are many practical ways to deliver quick and timely feedback to students. As you saw in figure 7.1 (page 165), a well-crafted feedback tool, like a rubric, will save you quite a bit of time while also providing students with specific points to reflect on and grow. Rubrics make the end goal obvious and measurable in a concrete way while quickening the process and making feedback discrete instead of ambiguous. Having the specific success criteria rubrics provide breaks down the larger learning targets, making it possible for learners to identify where to focus their efforts to improve overall. Likewise, when introducing a new concept, checking for understanding with a quick exit slip or entrance slip the next day and providing feedback with a simple + or – is also an efficient way of letting students know if they are on the right track.

There are limitless possibilities for effective feedback, so be creative in your feedback methods. Did you ever have a teacher who asked you and your student peers to trade papers as a means to quickly score an in-class quiz? Try this alternative: have students keep their quizzes and put all utensils away except for a pen that is a different color than the one they used to take the test. Then go over the correct answers, including the reasoning and evidence necessary to support why an answer is correct while having students adjust their own errors. Not only have you provided timely feedback, you were able to help students reflect and identify misconceptions and have also avoided the shame that some students feel when exposing their work to others. The benefit of timely feedback for students is clear—less time being confused and making the same errors over and over again. As this kind of feedback makes any student error trends become clear, follow this data gathering with a minilesson to refocus student learning and address those misconceptions. Even for those who demonstrate understanding on the first try, the minilesson will help to reinforce that understanding. Finally, as we cover next, collecting feedback gives you data to return to your team regarding any learning trends that you identified.

Performing Item Analysis

After an assessment, it is critical to not only provide students with timely feedback but also bring samples of student assessments back to your collaborative team for further discussion. In terms of analysis, your team should design assessments to gather data that are aligned with the team's learning targets and process standards. When doing this work during the assessment-design phase, there are two major lenses (perspectives) to employ while taking a deep dive into these data: (1) the whole class and (2) a sample pack—collections of a few assessments from different student populations. For example, your team might collect assessments based on students' reading levels, for students receiving specific forms of intervention, for students who are learning English, and so on.

When assessment structures align with a team's learning targets, gathering data for a whole class (or even multiple classes) is very easy to do, because they can be put together and aggregated. There are a number of tools that can support this process, such as Mastery Manager (www.masterymanager.com) and Google Classroom (https://classroom.google.com), and provide your team with a useful way to look for overarching trends of correct and incorrect responses.

The sample-pack method works best when your team wants to go deeper than just looking at whole-class trends, but rather seeks to zoom in on the progress of more specific student groups. Taking the time to thoroughly explore a subset of assessment results takes longer; however, it's an ideal approach for identifying where students may have made a wrong turn in the process, logic, or application.

When working with our team, we often employ both whole-class and sample-pack lenses. First, we look at final answer scores for an entire class or population, then also analyze a sample pack to further identify where students made specific mistakes. Consequently, these combined data guide our next steps of reteaching.

Figure 7.2 provides guiding questions for teams to conduct data analysis for both a whole class and via a sample pack.

1. Whole Class
   - As a whole, how did students perform? Are there any tasks or questions that proved problematic for the class as a whole? If yes, review how these topics or tasks were covered in your curriculum and teaching.
   - Do you see any general trends among student assessment data for multiple sections (classes) of the same course?
   - To what can you contribute student successes and challenges?
2. Sample Pack
   - What variations do you see in terms of performance by the group?
   - Is there a need for additional scaffolds for specific students demonstrating a common need?

Figure 7.2: Data-analysis perspectives.

*Visit **go.SolutionTree.com/literacy** for a free reproducible version of this figure.*

Once your team has collected and analyzed data, the next critical step is to decide what to do with what those data communicate. Different data points and data sets have different purposes. It may seem clear what to do after conducting an analysis of formative data, such as identifying where students need more instruction and using that knowledge to revisit a correlated learning target with a more scaffolded process. Benchmark or normative data, on the other hand, sometimes leave teams wondering what to do with them. In our collaborative team, we always start our discussion with the obvious: the fact that these data likely confirm or add evidence and clarification to what we have already observed of student proficiency. If this isn't the case, then we ask, What were the surprises? This task can be daunting, especially if the data are confirming that you may have readers in your class operating at a four-grade (or more) deficit from the content you are teaching. Don't let these data points thwart you; again, they only confirm what you probably already suspected. Even if they come as a surprise, the data are critical information for you and your team to have. As your team moves forward, these data give it the necessary marching orders for how to support students in developing a wide array of skills.

Teachers will not be able to craft a differentiated lesson for each student in the room, nor employ multiple texts and primary sources on the same topic each day. But, they can and should look for opportunities to do so throughout a unit, and social studies provides a great platform for this kind of work. We selected and crafted the strategies in chapters 3–6 to help you do just this.

As your team deploys these strategies, consider how creating student learning groups based on assessment data can be a very productive use of the data. Many online assessment systems, such as the assessment tools at Renaissance Learning (www.renaissance.com), have tools integrated as part of their reporting features that allow teams to easily group students and will even generate skill data for groups. Whether using an online data tool or old-school paper and pen, here are some basic grouping methods. Which one team members choose ultimately depends on the task and purpose.

▸ **Homogeneous groups:** Group students based on similar assessment results to focus on developing a group-specific skill or mastering a group-specific content goal.

▸ **Linear groups:** Create a list of students from highest to lowest scores and break them into groups accordingly.

▸ **Heterogeneous groups:** Group students with highly variable skill sets or proficiency levels into one group so that they can benefit from each other's areas of strength. The key to this grouping is to monitor each group's efforts to ensure students work as a community for mutual benefit rather than (as we sometimes see) one or two students working while the others are either lost or copying the work without understanding it. Another way to distribute groups heterogeneously is to list the students in a vertical row from highest score to lowest score. Then, cut the list in half, and place the two lists side-by-side. Each pair of names then works as a paired group.

While there are many more approaches your team could take, what's important is to think creatively about how you can group students in a way that allows them to grow and benefit from collaboration and collective strengths. While randomly grouping students by counting off in class or organizing based on birthdays may provide variety, these methods don't fit any specific purpose or further skill development and content mastery.

Wrapping Up

Meaningful assessment requires a great deal of intention and reflection from your team. While it is the students' responsibility to apply the skills they have developed, a team's role is to create assessments that intentionally and accurately mirror the curriculum and skills each team member teaches. This kind of intentional assessment and reflection lead to responsive teaching for the entire PLC. This act of inquiry fosters the thoughtful selection and development of tools, texts, and tasks that help students progress toward being critically literate and possessing the disciplinary literacy skills necessary to not only grow within your content classroom but also in the post–high school world.

Collaborative Considerations *for* Teams

- Does your team have a robust assessment cycle that balances formative and summative assessments?

- Does your team take student reading ability into account when selecting texts for use in assessments? Does the team have an awareness of text and task complexity?

- Has your team collaborated on how to best share feedback with students so that it will impact future learning and growth?

EPILOGUE

Our work has taught us that building capacity for collaboration among all teachers and literacy experts is truly the strongest catalyst for supporting student growth in every area of school curriculum. As schools commit to exemplary instruction to support the growth of every learner, literacy instruction within every discipline is something that educators with the power of a PLC behind them must not overlook. Regardless of how the teams in your PLC are structured, whether organized by discipline, by grade level, or vertically across departments, collaboration toward increasing literacy always begins with teaming teachers who are focused on improving their own capacity to impact student learning and scaffold critical literacy skills.

We hope this book allows teams in all areas of the social studies discipline to find guidance and strategies that are adaptable to each team member's classes. This book should be an easy resource available to all of you to meet the needs of your pre-, during-, and postreading and writing strategies. By adapting each strategy you use to meet the needs of all learners in your class, you and your team members can be creative, trying them in different capacities to achieve different learning goals. This is when we all learn best and learn more about our ability to achieve the mission of learning for *all* students.

While this book speaks directly to social studies experts who want to hone their teaching of literacy within their classroom, this is one installment in a series that will support teacher collaboration and strategic literacy infused teaching in all content areas. Each text in this series focuses on building a common language and literacy thought partners across disciplines. Ultimately, the literacy skills students develop in grades 6–12 will greatly impact their ability to think critically and their overall readiness for college and career.

APPENDIX:

REPRODUCIBLES

Reading Comprehension Process

Did I . . . ?	Strategic Comprehension Step	Before, During, or After Reading
☐	**Preview text**, ask questions, and make predictions.	**Before:** Focus and get ready to read.
☐	**Recall** what you already know about the topic.	
☐	Set a **purpose** for reading.	
☐	Make a **notetaking plan** for remembering what's important.	
☐	Define **key concepts** and important vocabulary whenever possible.	
☐	Keep your **purpose** for reading in mind.	**During:** Stay mentally active.
☐	**Make meaning** by: • Asking questions • Putting the main ideas into your own words • Visualizing what you read • Making notes to remember what's important • Making connections between the text and people, places, things, or ideas	
☐	**Be aware** of what's happening in your mind as you read. Consider: • Am I focused or distracted? • Do I need to go back to a part I didn't get and reread it? • What are my reactions to what I am reading?	
☐	**Reflect on** what you've read. Consider: • Did I find out what I needed or wanted to know? • Can I summarize the main ideas and important details in my own words? • Can I apply what I have learned? • Can I talk about or write about what I have learned?	**After:** Check for understanding.

Content-Standard Analysis Tool

Content Standard	Artifacts		Opportunities
	Lesson Plans and Course Materials	Student-Generated Evidence of Standard	Power Standard ☐
Unpacked			
	Strengths and Needs	Strengths and Needs	

Content Standard	Artifacts		Opportunities
	Lesson Plans and Course Materials	Student-Generated Evidence of Standard	Power Standard ☐
Unpacked			
	Strengths and Needs	Strengths and Needs	

Prereading Anticipation Guide

Name_____

Prereading Anticipation Guide for _____

Topic: _____

Before reading: Look at each statement carefully. Put a check in the appropriate column in Before Reading to indicate whether you agree or disagree with each statement.

During reading: In the center column, write the evidence from the reading that supports or contradicts the statement. Include the page number where you found your evidence.

After reading: Reread the statement and the evidence that supports or contradicts it. Put a check in the appropriate column in After Reading to indicate whether you now agree or disagree.

Before Reading		Statement	Page Number	After Reading	
Yes	No			Yes	No

Quick-Write Prompts

Reflect on the essential question.

For five minutes, write your thoughts about the following question.

Reflect on the essential question.

For five minutes, write your thoughts about the following question.

Reflect on the essential question.

For five minutes, write your thoughts about the following question.

Frayer Model Template

Directions: Write down a vocabulary word or concept that was critical to your reading and explain it using the included prompts.

Definition (what it is):	Characteristics:

Vocabulary Word:

Examples or pictures:	What it is not:

Reading and Writing Strategies for the Secondary Social Studies Classroom in a PLC at Work © 2021 Solution Tree Press

SolutionTree.com • Visit **go.SolutionTree.com/literacy** to download this free reproducible.

Beliefs and Opinions Survey

Directions: Read the following statements and circle your reactions. Be prepared to defend your opinions with evidence from your own experiences.

Statement:				
Strongly Disagree	Disagree	Neutral	Agree	Strongly Agree

Statement:				
Strongly Disagree	Disagree	Neutral	Agree	Strongly Agree

Statement:				
Strongly Disagree	Disagree	Neutral	Agree	Strongly Agree

Statement:				
Strongly Disagree	Disagree	Neutral	Agree	Strongly Agree

Statement:				
Strongly Disagree	Disagree	Neutral	Agree	Strongly Agree

Statement:				
Strongly Disagree	Disagree	Neutral	Agree	Strongly Agree

Statement:				
Strongly Disagree	Disagree	Neutral	Agree	Strongly Agree

Find Evidence to Support Claims

Find evidence of the following.		
Thematic Claim Statement	**Is Claim Accurate?**	**Evidence to Support or Refute the Statement (include page numbers)**

Text-Dependent Questioning Graphic Organizer

Text or topic:

Question Category	Questioning the Author or Document General Questions	My Focused Questions
Said what? What is the author or document saying?	What is the author or document telling you? What does the author or document say that you need to clarify? What can you do to clarify what the author or document says? What does the author or document assume you already know?	
Did what? What did the author or document do?	How does the author or document tell you the information? Why is the author or document telling (or showing) you about this event, statistic, era, example, or visual? What does the vocabulary reveal about the author, content, or document? How does the author or document signal what is most important? How does the author construct his or her text or develop his or her ideas?	

Reading and Writing Strategies for the Secondary Social Studies Classroom in a PLC at Work © 2021 Solution Tree Press
SolutionTree.com • Visit **go.SolutionTree.com/literacy** to download this free reproducible.

Question Category	Questioning the Author or Document General Questions	My Focused Questions
So what? What might the author or document mean?	What does the author or document want you to understand? Why is the author or document telling you this? Does the author or document explain why something is so? What point is the author or document making here? What is the author's or document's purpose, and what support (evidence or reasoning) does the author or document present?	
Now what? What can you do with your understanding of the author or document?	How does this connect or apply to what I know? How does what the author or document says influence or change my thinking? What implications can I draw from what the author or document has told me?	

Source: Adapted from Buehl, 2017.

Primary Source Graphic Organizer

Prereading	
Type of Document Is this a letter, an article, an advertisement, a government document, or something else?	**Author of Document** Who wrote the document? What prior knowledge do you have about the author?
Date of Document When was the document created? What connects the document to other events or people during this time?	**Source of Document** In what city, town, or country was the document created? Where was it originally published? Do you have prior knowledge about the source?

During Reading	
Style of Document What is unusual about the language used in the document?	**Point of View of Document** Is the document written in the first person, using the pronoun I? How does the point of view affect the reader?
Vocabulary of Document What are key vocabulary words in the document? What words are challenging to understand?	

page 1 of 2

Postreading	
Main Idea of Document What is the main point the writer is presenting? Summarize the entire document in two or three sentences with a focus on the main ideas.	**Impact of Document** What feelings does the document bring up in you? How does it connect to our unit of study?
Questions Raised by Document What do you want to know more about? What is still unknown?	**Further Research** What are some other documents worth finding and reading?

Five Words Recording Sheet

Directions: While reading the assigned text, underline key words and phrases. After reading, choose the five most important words from the reading and add them to the Individual Selections column. When instructed, as a small group, discuss the words in the Individual Selections column and come to a consensus on the five most important words that the group agrees on. Add those words to the Group Consensus column.

Individual Selections	Group Consensus
1.	1.
2.	2.
3.	3.
4.	4.
5.	5.

Discussion Questions

Which words can your group agree upon?

Which words led to disagreements?

How did your thinking change as a result of your discussion?

Student-Generated Questioning Taxonomy

Directions:

Level of Thinking	Comprehension Self-Assessment	Focusing Questions
Creating	I have created new knowledge about the past.	
Evaluating	I can critically examine this author's arguments about the past.	
Analyzing	I can understand why this piece was created by placing it in a historical context.	
Applying	I can use my understanding to better comprehend how the past influences my life and world.	
Understanding	I can understand what the author is telling me about the past.	
Remembering	I can recall specific details, information, and ideas from this text.	

3–2–1 Activity

3. Identify three key details from the reading.

2. Write down two connections to other texts we have explored in this class.

1. Write down one question you are wondering about after reading.

1.	1.	1.
2.	2.	
3.		

Two-Quote Paragraph Template

Sentence Number	Type of Sentence
1	**Topic sentence:** State your claim (purpose of the text).
2–3	**Lead-in:** Establish a context for the upcoming evidence or explain a little about your claim. You might define vocabulary or add context to further your idea.
4	**Quotation or evidence:** Make sure you choose one of the best quotes or pieces of evidence from the text that will support your claim.
5–6	**Explanation or reasoning:** Now explain why your evidence is important. How and why does this evidence support your claim? Explain the reasoning behind why your evidence proves your claim.

Reading and Writing Strategies for the Secondary Social Studies Classroom in a PLC at Work © 2021 Solution Tree Press
SolutionTree.com • Visit **go.SolutionTree.com/literacy** to download this free reproducible.

7	**Quotation or evidence:** Make sure your second quote or piece of evidence supports your claim as well. Make sure you transition carefully into the evidence—don't just drop it in the middle of the paragraph with no sentence setting it up.
8–9	**Explanation or reasoning:** Now explain why your evidence is important. How does it support your claim? Explain the reasoning why your evidence proves your claim.
10	**Return to the claim (concluding sentence):** This last sentence should reconnect your evidence and ideas back to your primary claim.

Reading and Writing Strategies for the Secondary Social Studies Classroom in a PLC at Work © 2021 Solution Tree Press
SolutionTree.com • Visit **go.SolutionTree.com/literacy** to download this free reproducible.

Evaluating Claim Statements and Supporting Them With Evidence

Question: _____

After analyzing the text, three students wrote claims to answer the preceding question. Read each of their claims. Put a checkmark next to the student you feel has the best claim. After choosing a student's claim, reread the document and find at least ____ pieces of evidence to support the claim you selected.

_____ **Claim 1:** _____.

_____ **Claim 2:** _____.

_____ **Claim 3:** _____.

Evidence	Page Number	How the Evidence Supports Your Chosen Claim

Rubric to Evaluate a Writing Sample

Subject: _____

Student name: _____

Directions: Read and reflect on the following essential question. Then, read each sample claim statement and body paragraph. Next, assess the body paragraphs using the Evidence and Supporting Argument portion of the embedded rubrics. Finally, beneath each rubric, explain the score using the language of the rubric.

Essential question:

Sample A (claim):

Body paragraph:

Reading and Writing Strategies for the Secondary Social Studies Classroom in a PLC at Work © 2021 Solution Tree Press
SolutionTree.com • Visit **go.SolutionTree.com/literacy** to download this free reproducible.

	Scoring Criteria			
Writing Criteria	**1—Needs Improvement**	**2—Approaching Mastery**	**3—Mastery**	**4—Exceeds Mastery**
Historical Evidence (Factual Information)	Inappropriate or no use of evidence	Makes little use of evidence or information is inaccurate or irrelevant	Clearly states and generally defines evidence to support the claim (argument)	Clearly states and thoroughly defines evidence to support the claim
Supporting Argument (Analysis)	Inadequate or inaccurate analysis of the evidence	Introduces evidence, but analysis may be limited or simplistic	Includes evidence and, in most cases, explains how it supports the claim	Includes evidence that always specifically and consistently explains how it supports the claim

Explanation:

Reading and Writing Strategies for the Secondary Social Studies Classroom in a PLC at Work © 2021 Solution Tree Press
SolutionTree.com • Visit **go.SolutionTree.com/literacy** to download this free reproducible.

Historical Sentence Starters

Cause and Effect

There were economic, political, and social factors that contributed to . . .

_____ led to _____. . .

The most likely reason that . . .

_____ had a significant impact on _____ because . . .

_____ was a result of _____. . .

Several factors contributed to _____, such as . . .

Interpretation or Perspective

These actions changed the course of _____ because . . .

This was a significant event because . . .

One interpretation of this is. . . .

Others argue that . . .

The evidence suggests . . .

One way to interpret this event is . . .

Examining the event through the perspective of . . .

Sourcing

The author may be motivated by . . .

Even though the author says _____, he or she may mean . . .

Based on the era of this source . . .

At the time of this source, _____ was occurring, so . . .

The author may be biased because . . .

REFERENCES AND RESOURCES

Abosalem, Y. (2016, March 6). Assessment techniques and students' higher-order thinking skills. *International Journal of Secondary Education*, *4*(1), 1–11. Accessed at https://pdfs.semanticscholar.org/81e4/0f2f1321180b6acf5de0d53b7f05251ba030 .pdf on August 12, 2019.

Allison, B., & Rehm, M. (2007). Effective teaching strategies for middle school learners in multicultural, multilingual classrooms. *Middle School Journal*, *39*(2), 12–18. Accessed at www.jstor.org/stable/23048334 on July 3, 2020.

Anderson, L. W., & Krathwohl, D. (Eds.). (2001). *A taxonomy for learning, teaching, and assessing: A revision of Bloom's taxonomy of educational objectives.* New York: Longman.

Aronson, E., & Patnoe, S. (1997). *The jigsaw classroom: Building cooperation in the classroom (2nd ed.).* New York: Addison Wesley Longman.

Bailey, K. & Jakicic, C. (2012). *Common formative assessment: A toolkit for Professional Learning Communities at Work.* Bloomington, IN: Solution Tree Press.

Urquhart, V., & Frazee, D. (2012). *Teaching reading in the content areas: If not me, then who?* (3rd ed.). Alexandria, VA: Association for Supervision and Curriculum Development.

Blanton, W. E., Wood, K. D., & Moorman, G. B. (1991, April). The role of purpose in reading. *Education Digest*, *56*(8), 33.

Bloom, B. S. (Ed.). (1956). *Taxonomy of educational objectives, handbook 1: The cognitive domain.* New York: Longman.

Buehl, D. (2017). *Developing readers in the academic disciplines* (2nd ed.). Portland, ME: Stenhouse.

Buffum, A., Mattos, M., & Malone, J. (2018). *Taking action: A handbook for RTI at Work.* Bloomington, IN: Solution Tree Press.

Bunting, E., & Diaz, D. (1994). *Smoky night.* San Diego, CA: Harcourt Brace.

Carruthers & Batsford. (Publisher). (1916). The Allies—"Onward to victory." Winnipeg, Canada. Accessed at https://commons.wikimedia.org/wiki/File:%22Onward_to _Victory%22,_World_War_I_Allied_propaganda_postcard.jpg on June 2, 2020.

Center for Comprehensive School Reform and Improvement. (2007). *A teacher's guide to differentiating instruction.* Accessed at https://files.eric.ed.gov/fulltext/ED495740.pdf on January 9, 2019.

Columbia University Teachers College. (2005). The academic achievement gap: Facts & figures. Accessed at www.tc.columbia.edu/articles/2005/june/the-academic -achievement-gap-facts--figures on June 7, 2020.

Conzemius, A. E., & O'Neill, J. (2014). *The handbook for SMART school teams: Revitalizing best practices for collaboration.* Bloomington, IN: Solution Tree Press.

Cook, B. W. (1999). *Eleanor Roosevelt: Volume 2, the defining years, 1933–1938.* New York: Penguin.

CRAAP Test. (n.d.). In *Wikipedia.* Accessed at https://en.wikipedia.org/wiki/CRAAP _test on May 12, 2020.

Crawford, J. (2011). *Using power standards to build an aligned curriculum: A process manual.* Thousand Oaks, CA: Corwin.

Csikszentmihalyi, M. (2009). *Flow: The psychology of optimal experience.* New York: HarperCollins.

de Koning, B., & van der Schoot, M. (2013). Becoming part of the story! Refueling the interest in visualization strategies for reading comprehension. *Educational Psychology Review, 25*(2), 261–287.

DuFour, R. (2004). What is a "professional learning community"? *Educational Leadership, 61*(8), 6–11. Accessed at www.siprep.org/uploaded/ProfessionalDevelopment /Readings/PLC.pdf on May 28, 2019.

DuFour, R., DuFour, R., Eaker, R., Many, T.W., & Mattos, M. (2016). *Learning by doing: A handbook for Professional Learning Communities at Work* (3rd ed.). Bloomington, IN: Solution Tree Press.

Eglash, R. (2019). "Hi! My name's Eva": A teenage Holocaust victim's diary comes to life on Instagram. *Washington Post.* Accessed at www.washingtonpost.com/world/2019 /05/02/hi-my-names-eva-teenage-holocaust-victims-diary-comes-life-instagram /?noredirect=on&utm_term=.bba6fee5d6ea on May 22, 2020.

Ferlazzo, L., & Sypnieski, K. H. (2018). *Activating prior knowledge with English language learners.* Accessed at www.edutopia.org/article/activating-prior-knowledge-english -language-learners on August 26, 2020.

Fountas, I. C., & Pinnell, G. S. (2012). *Genre study: Teaching with fiction and nonfiction books.* Portsmouth, NH: Heinemann.

Frayer, D. A., Frederick, W. C., & Klausmeier, H. J. (1969). *A schema for testing the level of cognitive mastery.* Madison, WI: Wisconsin Center for Education Research.

Gabriel, R., & Wenz, C. (2017). Three directions for disciplinary literacy. *Educational Leadership, 74*(5). Accessed at www.ascd.org/publications/educational-leadership /feb17/vol74/num05/Three-Directions-for-Disciplinary-Literacy.aspx on May 28, 2019.

Gallagher, K. (2018, February 27). Mass shooting unit: Day 5. Accessed at www
.kellygallagher.org/kellys-blog/mass-shooting-unit-day-5 on May 28, 2019.

Gambrell, L. B. (2011). Seven rules of engagement: What's most important to know
about motivation to read. *The Reading Teacher, 65*(3), 172–178. Accessed at www
.readinghalloffame.org/sites/default/files/final_pdf_of_ms_10.1002_trtr.01024.pdf
on June 8, 2020.

Good, M. E. (2006). *Differentiated instruction: Principles and techniques for the
elementary grades.* Unpublished doctoral dissertation, Dominican University of
California, San Rafael, CA.

Goodwin, B. (2017). Research matters: Helping students develop schemas. *Educational
Leadership, 75*(2), 81–82.

Iamarino, D. L. (2014). The benefits of standards-based grading: A critical evaluation of
modern grading practices. *Current Issues in Education, 17*(2). Accessed at http://cie
.asu.edu/ojs/index.php/cieatasu/article/view/1234 on July 20, 2020.

Inaim, R. (n.d.). Bloom's taxonomy. Accessed at https://kpu.pressbooks.pub/app/uploads
/sites/4/2018/06/Blooms-Graphic.png on June 8, 2020.

International Society for Technology in Education. (n.d.). *ISTE standards for students.*
Accessed at www.iste.org/standards/for-students on September 3, 2019.

Live Election Results: Presidential Election. (2016). *Washington Post.* Accessed at www
.washingtonpost.com/2016-election-results/us-presidential-race/?utm_term=
.5dbf7236f1ee on May 22, 2020.

Marzano, R. J. (2017). *The new art and science of teaching.* Bloomington, IN: Solution
Tree Press.

McFarland, J., Hussar, B., Wang, X., Zhang, J., Wang, K., Rathbun, A., et al. (2018).
The condition of education 2018. Accessed at https://nces.ed.gov/pubs2018/2018144
.pdf on May 29, 2019.

McKnight, K. S. (2010). *The teacher's big book of graphic organizers: 100 reproducible
organizers that help kids with reading, writing, and the content areas.* San Francisco:
Jossey-Bass.

National Center on Intensive Intervention. (n.d.). *Intensive intervention & multi-tiered
systems of support (MTSS).* Accessed at https://intensiveintervention.org/intensive
-intervention/multi-tiered-systems-support on September 4, 2019.

National Council for the Social Studies. (2017). *The College, Career, and Civic Life (C3)
Framework for Social Studies State Standards: Guidance for enhancing the rigor of
K–12 civics, economics, geography, and history.* Silver Spring, MD: Author. Accessed
at www.socialstudies.org/sites/default/files/2017/Jun/c3-framework-for-social
-studies-rev0617.pdf on April 27, 2020.

National Governors Association Center for Best Practices & Council of Chief State
School Officers. (2010). *Common Core State Standards for English language arts and
literacy in history/social studies, science, and technical subjects.* Accessed at www
.corestandards.org/assets/CCSSI_ELA%20Standards.pdf on May 28, 2019.

Pardede, P. (2017). *A review on reading theories and its implication to the teaching of reading*. Paper presented at the Bimonthly Collegiate Forum of Universitas Kristen Indonesia, Jakarta, Indonesia.

Paul, K. (2010, January 11). Why wars no longer end with winners and losers. *Newsweek*. Accessed at www.newsweek.com/why-wars-no-longer-end-winners-and-losers-70865 on April 28, 2020.

Pearcy, T., & Dickson, M. (Eds.). (1997). *Reply of the House of Commons to King James I*. Accessed at https://wwnorton.com/college/history/ralph/workbook/ralprs20a.htm on July 6, 2020.

ProLiteracy. (n.d.). *U.S. adult literacy facts*. Accessed at https://proliteracy.org/Portals/0 /pdf/PL_AdultLitFacts_US_flyer.pdf?ver=2016-05-06-145137-067 on September 8, 2020.

Rebell, M. A. (2008, February). Equal opportunity and the courts. *Phi Delta Kappan*, *89*(6), 432–439.

Roberts, S., & Bubar, J. (2020, January 6). *Sitting down to take a stand*. Accessed at https://upfront.scholastic.com/issues/2019-20/010620/sitting-down-to-take-a -stand.html on May 22, 2020.

RTI Action Network. (n.d.). *Tiered instruction/intervention*. Accessed at http:// rtinetwork.org/essential/tieredinstruction on January 18, 2019.

Rumelhart, D. E. (1980). Schemata: The building blocks of cognition. In R. J. Spiro, B. C. Bruce, & W. F. Brewer (Eds.), *Theoretical issues in reading comprehension: Perspective and cognitive psychology, linguistics, artificial intelligence, and education* (pp. 33–58). Hillsdale, NJ: Erlbaum.

Shanahan, T., & Shanahan, C. (2008). Teaching disciplinary literacy to adolescents: Rethinking content-area literacy. *Harvard Educational Review*, *78*(1), 40–59, 279. Accessed at https://dpi.wi.gov/sites/default/files/imce/cal/pdf/teaching-dl.pdf on January 17, 2019.

Shi, D. E., & Tindall, G. B. (2016). *America: a narrative history* (10th ed.). New York: W. W. Norton & Co.

Stephens, D., Morgan, D. N., DeFord, D. E., Donnelly, A., Hamel, E., Keith, K. J., et al. (2011). The impact of literacy coaches on teachers' beliefs and practices. *Journal of Literacy Research*, *43*(3), 215–249.

Stiggins, R. (2005). From formative assessment to assessment FOR learning: A path to success in standards-based schools. *Phi Delta Kappan*, *87*(4), 324–328.

ThinkCERCA. (n.d.a). *Our story*. Accessed at www.thinkcerca.com/story on May 18, 2020.

ThinkCERCA. (n.d.b). *What is CERCA? Equip students for the real world with a research-based literacy framework*. Accessed at www.thinkcerca.com/cerca on May 18, 2020.

Tovani, C. (2000). *I read it, but I don't get it: Comprehension strategies for adolescent readers*. Portland, ME: Stenhouse.

Tovani, C. (2005). The power of purposeful reading. *Educational Leadership*, *63*(2), 48–51.

Townsley, M. & Wear, N. L. (2020). *Making grades matter: Standards-based grading in a Secondary PLC at Work*. Bloomington, IN: Solution Tree Press.

U.S. Department of Education, National Center for Education Statistics, National Assessment of Educational Progress. (n.d.). *NAEP reading report card*. Accessed at www.nationsreportcard.gov/reading_2017?grade=4 on September 4, 2019.

U.S. Department of Education, National Center for Education Statistics, National Assessment of Educational Progress. (2016). *Reading performance*. Accessed at https://nces.ed.gov/programs/coe/pdf/Indicator_CNB/coe_cnb_2016_05.pdf on March 12, 2019.

USHistory.org. (2016). *Witchcraft in Salem*. Accessed at www.commonlit.org/en/texts/witchcraft-in-salem on June 8, 2020.

Watanabe-Crockett, L. (2019). *Future-focused learning: 10 essential shifts of everyday practice*. Bloomington, IN: Solution Tree Press.

Wilhelm, J. D. (2007). *Engaging readers and writers with inquiry: Promoting deep understandings in language arts and the content areas with guiding questions*. New York: Scholastic.

Wineburg, S., & McGrew, S. (2017). *Lateral reading: reading less and learning more when evaluating digital information*. Accessed at https://papers.ssrn.com/sol3/papers.cfm?abstract_id=3048994 on July 6, 2020.

Wineburg, S., & McGrew, S. (2018). *Lateral reading and the nature of expertise: Reading less and learning more when evaluating digital information*. Accessed at https://purl.stanford.edu/yk133ht8603 on June 8, 2020.

Worthy, J., & Broaddus, K. (2001). Fluency beyond the primary grades: From group performance to silent, independent reading. *Reading Teacher, 55*(4), 334–343.

INDEX

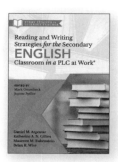

Reading and Writing Strategies for the Secondary English Classroom in a PLC at Work®
Edited by Mark Onuscheck and Jeanne Spiller
Close literacy achievement gaps across grades 6–12. Part of the Every Teacher Is a Literacy Teacher series, this resource highlights how English language arts teachers can work collaboratively to combat literacy concerns and improve student skill development.
BKF904

Reading and Writing Strategies for the Secondary Science Classroom in a PLC at Work®
Edited by Mark Onuscheck and Jeanne Spiller
Equip your students with the literacy support they need to think like scientists. Written by a team of experienced educators, this book provides practical literacy-based strategies specifically for science teachers of grades 6–12.
BKF907

The New Art and Science of Teaching Reading
Julia A. Simms and Robert J. Marzano
The New Art and Science of Teaching Reading presents a compelling model for reading development structured around five key topic areas. More than 100 reading-focused instructional strategies are laid out in detail to help teachers ensure every student becomes a proficient reader.
BKF811

The New Art and Science of Teaching Writing
Kathy Tuchman Glass and Robert J. Marzano
Using a clear and well-organized structure, the authors apply the strategies originally laid out in *The New Art and Science of Teaching* to the teaching of writing. In total, the book explores more than 100 strategies for teaching writing across grade levels and subject areas.
BKF796

"Tremendous, tremendous, tremendous!

The speaker made me do some very deep internal reflection about the **PLC process** and the personal responsibility I have in making the school improvement process work **for ALL kids**."

—Marc Rodriguez, teacher effectiveness coach, Denver Public Schools, Colorado